# Radiation Measurement in Photobiology

# Biological Techniques Series

J. E. TREHERNE
*Department of Zoology*
*University of Cambridge*
*England*

G. EVAN
*The Ludwig Institute for*
*Cancer Research*
*MRC Centre, Cambridge*
*England*

Ion-sensitive Intracellular Microelectrodes, *R. C. Thomas*, 1978
Time-lapse Cinemicroscopy, *P. N. Riddle*, 1979
Immunochemical Methods in the Biological Sciences: Enzymes and
    Proteins, *R. J. Mayer* and *J. H. Walker*, 1980.
Microclimate Measurement for Ecologists, *D. W. Unwin*, 1980
Whole-body Autoradiography, *C. G. Curtis, S. A. M. Cross,*
    *R. J. McCulloch* and *G. M. Powell*, 1981
Microelectrode Methods for Intracellular Recording and Ionophoresis,
    *R. D. Purves*, 1981
Red Cell Membranes—A Methodological Approach, *J. C. Ellory* and
    *J. D. Young*, 1982
Techniques of Flavonoid Identification, *K. R. Markham*, 1982
Techniques of Calcium Research, *M. V. Thomas*, 1982
Isolation of Membranes and Organelles from Plant Cells, *J. L. Hall* and
    *A. L. Moore*, 1983
Intracellular Staining of Mammalian Neurones, *A. G. Brown* and
    *R. E. W. Fyffe*, 1984
Techniques in Photomorphogenesis, *H. Smith* and *M. G. Holmes*, 1984
Principles and Practice of Plant Hormone Analysis, *L. Rivier* and
    *A. Crozier*, 1987
Wildlife Radio Tagging, *R. Kenward*, 1987
Immunochemical Methods in Cell and Molecular Biology, *R. J. Mayer*
    and *J. H. Walker*, 1987
Radiation Measurement in Photobiology, *B. L. Diffey*, 1989

# Radiation Measurement in Photobiology

*Edited by*

B. L. DIFFEY

*Regional Medical Physics Department*
*Durham Unit*
*Dryburn Hospital*
*Durham DH1 5TW, UK*

ACADEMIC PRESS

*Harcourt Brace Jovanovich, Publishers*

London    San Diego    New York    Berkeley
Boston    Sydney    Tokyo    Toronto

ACADEMIC PRESS LIMITED
24–28 Oval Road
London NW1 7DX

*U.S. Edition published by*
ACADEMIC PRESS INC.
San Diego, CA 92101

**British Library Cataloguing in Publication Data**
Radiation measurement in photobiology.
  1. Photobiology
  I. Diffey, B.L.   II. Series
  574.19′153

ISBN 0-12-215840-7

Typeset by Mathematical Composition Setters Ltd, Salisbury, Wilts
Printed in Great Britain by T. J. Press (Padstow) Ltd, Padstow, Cornwall

# List of Contributors

L. O. BJÖRN, *Department of Plant Physiology, University of Lund, Box 7007, S-220 07 Lund, Sweden*

B. L. DIFFEY, *Regional Medical Physics Department, Dryburn Hospital, Durham DH1 5TW, UK*

P. GIBSON, *Glen Spectra Ltd, 2–4 Wigton Gardens, Stanmore, Middlesex HA7 1BG, UK*

T. M. GOODMAN, *National Physical Laboratory, Teddington, Middlesex TW11 0LW, UK*

M. G. HOLMES, *Department of Botany, University of Cambridge, Downing Street, Cambridge CB2 3EA, UK*

D. PHILLIPS, *The Royal Institution, 21 Albemarle Street, London W1X 4BS, UK*

M. SEYFRIED, *Universität Karlsruhe, Botanisches Institut 1, Kaiserstrasse 12, D–7500 Karlsruhe 1, FRG*

A. W. S. TARRANT, *Department of Chemical & Process Engineering, Home Economics & Domestic Engineering Research Unit, University of Surrey, Guildford, Surrey GU2 5XH, UK*

A. D. WILSON, *Applied Physics Group, Pilkington Optronics, Barr & Stroud Ltd, Caxton Street, Anniesland, Glasgow G13 1HZ, UK*

# Contents

# Preface

It is with pleasure that I write a preface to introduce this book on the important topic of the measurement of optical radiation and its application in photobiology. This volume arose out of a meeting of the British Photobiology Society held in the historic buildings of the Royal Institution, London, and was organized by Dr Brian Diffey.

The first three chapters are concerned with fundamental notions and definitions, optical radiation detectors based on physical principles, and the problems associated with calibration.

The next three chapters deal with important applications and extensions of these radiant measurements, including a short review of biological and medical users of lasers. As a dermatologist I admired and envied the rigorous standards a botanist can apply to assessing action spectra in plants.

The final three chapters on specialized studies and developments illustrate well the wide diversity that exists in photobiology. These cover ultraviolet radiation dosimetry using polymer films, computer modelling of terrestrial ultraviolet radiation and the "diffusion optics" in biological media.

Clearly the necessity of quantifying stimuli and responses is most important in all branches of biology, particularly in photobiology. This book I hope will stimulate interest and foster the best standards.

I. A. MAGNUS

*Institute of Dermatology*
*London, UK*

# 1
# Basic Principles of Light Measurement

## A. W. S. TARRANT

*Department of Chemical and Process Engineering*
*Home Economics and Domestic Engineering Research Unit*
*University of Surrey*
*Guildford*
*Surrey GU2 5XH, UK*

## 1.1. Light and Radiation — Introduction

Logically, one should talk about "radiation" first and then "light" as a special case of it. But in this chapter I have chosen to discuss "light" first. That is because we can readily visualize "light"; it seems simpler to picture the various concepts in terms of light that we can see, and familiar lamps that produce it, than to work in terms of unseen "radiation". To avoid duplication the basic concepts are dealt with under the heading of "light", so it is important that the reader who is not concerned with visible light should not skip sections of this chapter.

This book is intended for biological scientists unfamiliar with mathematical and physical concepts. This chapter assumes no mathematical pre-knowledge of the reader, and I must ask my colleagues of those disciplines to bear with a lot of words rather than a few equations.

### 1.1.1. What do we mean by "light"?

Before we can measure light properly we have to be quite certain about what we mean by "light". In common conversation we often use that word in a very loose way, sometimes to the extent of talking about "ultraviolet light" when what we mean is radiation that is invisible to the eye. So in scientific terms what do we mean by "light"?

Imagine that we have a beam of light coming out of, say, a slide projector (Fig. 1.1). For the time being let us assume that there is no slide in the projector, so were we to shine this beam on a projection screen we should just see a plain patch of white light. White light in fact is made up of light of a whole variety of wavelengths, and if we make the light of different

RADIATION MEASUREMENT IN PHOTOBIOLOGY
ISBN 0-12-215840-7

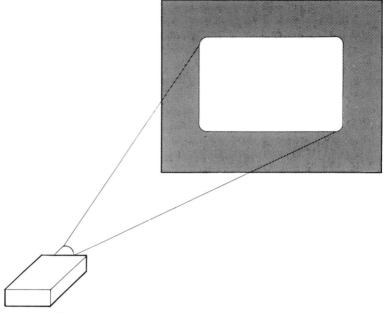

**Fig. 1.1**   A projector producing a defined beam of light.

wavelengths go in different directions by putting a prism in front of the projector then we shall see that light spread out into a spectrum (Fig. 1.2). The spectrum that we get will not appear equally bright in all parts. With an ordinary slide projector the brightest part will usually be in the yellowish-green part; as you go towards the red—longer wavelengths—it will become less bright and ultimately will fade out altogether. If you go towards the blue end—shorter wavelengths—from the yellow part it will get less bright until it fades out in the violet. Now if we examine the part beyond the red end we find in fact that there is radiation coming out of the projector and falling there; we cannot see it, but we can detect it with physical detectors such as silicon photodiodes. There may even be enough radiation there for us to actually feel it—by sensing the heat it produces—on our hands. We call this radiation "infra-red" radiation. Likewise if we look beyond the violet part we find radiation coming out of our projector that we cannot see. We can photograph it, and demonstrate its existence by making it cause fluorescence.

If we could find out how much energy was coming out of our projector at different wavelengths we would find it as shown in Fig. 1.3, with very much more energy in the red and infra-red than in the blue or ultraviolet. Strictly

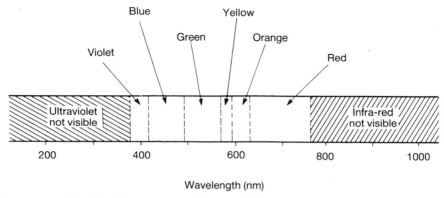

Fig. 1.2 The spectrum, with a rough indication of the colours seen.

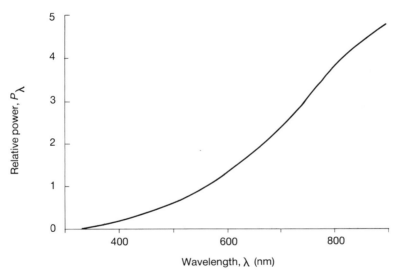

Fig. 1.3 The relative spectral distribution of power in the visible region from a typical projector incorporating a tungsten halogen lamp.

we should speak in terms of "power" coming out of our projector rather than "energy", because we are concerned with the rate at which energy comes out rather than energy as such. How is it then that the yellow or green seems to be the brightest part?

The fact is that our eyes are differently sensitive to different wavelengths —we are most sensitive to yellow-green light; less so to red and blue, and not at all to infra-red and ultraviolet. Obviously if we are going to measure "light" in physical terms we have to take this factor into account.

The relative sensitivity of the human eye to radiation of various wavelengths has been much studied over the years. The usual method of study involves asking human subjects to view a field in an optical instrument of which one-half is illuminated with light of a known single wavelength, whilst the other is illuminated with white light. The observer is asked to adjust the brightness until they appear to be equally bright. This is repeated for a series of different wavelengths throughout the spectrum, and a set of mean results compiled for a group of several observers. The sensitivity of any wavelength $\lambda$, relative to that of the maximum sensitivity at wavelength $\lambda_{max}$, is given by the inverse ratio of the amount required to match a constant white at $\lambda$ and $\lambda_{max}$. For example, if at a certain wavelength in the orange, say 610 nm, it requires 6 times as much power to match the white as for the yellow-green of $\lambda_{max}$, then the eye sensitivity to radiation of wavelength 610 nm is obviously one-sixth of that of the maximum.

In this way the curve representing the relative spectral sensitivity curve of the human eye can be determined. It is usually spoken of as "the visibility function", but it is officially called the "spectral luminous efficiency curve". In some older books it is referred to as the "relative luminous efficiency" curve. Notice that it is indeed only a **relative** sensitivity curve; it is only the ratio of the sensitivity at any wavelength to that of the maximum.

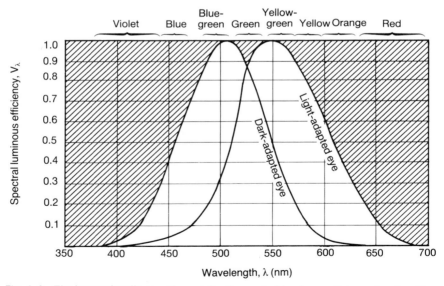

**Fig. 1.4**  The internationally agreed curve for the spectral luminous efficiency function (the "visibility function"). Reproduced from Henderson and Marsden (1972).

In practice the curves obtained by different individuals with normal vision are closely similar. The standard curve is illustrated in Fig. 1.4, and the internationally agreed standard data can be found in the British Standards Publication BS 4727 part 2, and in any textbook on photometry (e.g. Henderson and Marsden, 1972). The curve shown is that for normal (photopic) vision. A different curve, also shown in Fig. 1.4, applies to the dark-adapted eye (scotopic) vision.

## 1.1.2. The measurement of light

### 1.1.2.1. Luminous flux

Fortunately for our purposes the human eye in normal vision operates in a strictly additive way; if it is presented with radiation of several wavelengths simultaneously the response is simply the sum of the individual responses to radiation of each wavelength concerned. We can use, therefore, the spectral luminous efficiency curve to quantify "light"—that is our visual sensation —provided we know how much power at each wavelength is emitted from our source of radiation. If we divide that power spectrum into narrow strips, as in Fig. 1.5, we can calculate the light-producing effect of each narrow waveband, and add them up to obtain a measure of the total amount of "light". Exactly how this is done is explained in the following

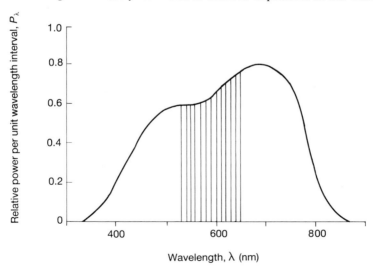

**Fig. 1.5** The spectral power distribution of a light source divided into a series of narrow wavebands.

section, but note here that we are getting involved in psychophysical measurements—we are introducing psychophysical data in the shape of the spectral luminous efficiency curve to the purely physical process of light measurement.

Consider again our projector of Fig. 1.1. We shall consider first the whole amount of light in the beam, which will be the same whether we examine it close to the projector where it occupies a small area, or nearer the screen where it extends over a larger area; the **amount** of light is the same.

To obtain a quantitative measure, we shall need to know also the power in the beam at each wavelength; let us suppose that $P_\lambda$ represents the power (in watts) per unit wavelength interval. The power between two wavelengths which are slightly different, say $\lambda$ and $\lambda + \delta\lambda$, will then be $P_\lambda \, \delta\lambda$. With most light sources $P_\lambda$ will be different for every different wavelength, as in Fig. 1.3. Now as far as the light-producing effect is concerned, the effect of that power between $\lambda$ and $\lambda + \delta\lambda$ will be $P_\lambda \, \delta\lambda$ multiplied by the appropriate relative sensitivity value $V_\lambda$, i.e. $P_\lambda V_\lambda \, \delta\lambda$. If we now think not of just that one narrow waveband, but of the whole spectrum, then we will simply have to add up the $P_\lambda V_\lambda \, \delta\lambda$ values over the whole spectrum as in Fig. 1.5. We write the sum as $\Sigma P_\lambda V_\lambda \, \delta\lambda$ where $\Sigma$ means "the sum of". To put this in formal mathematical language this sum becomes

$$\int_0^\infty P_\lambda V_\lambda \, d\lambda$$

which means the sum of all the $P_\lambda V_\lambda$ values that you would get between zero and infinity wavelengths if you made the steps of Fig. 1.5 (which are $\delta\lambda$ in width) infinitesimally small. But for all practical purposes this is the same thing as $\Sigma P_\lambda V_\lambda \, \delta\lambda$.

Now this gives us a number of light units. In practice light units had been used for the best part of 100 years before SI units came into being, and to save changing the established units we multiply the "units" that we have just arrived at by a constant to make them match the existing units. We call this constant "$k_m$" so the number of light units in our projector beam is now:

$$k_m \, \Sigma \, P_\lambda V_\lambda \, \delta\lambda$$

or

$$k_m \int_0^\infty P_\lambda V_\lambda \, d\lambda.$$

What we have done is to weight the power in our projector beam by a factor depending on the light-producing effect of each wavelength, and the figure that we have arrived at is a measure of the visible light in our beam. We call this "amount of light" the **luminous flux** in our beam, and the unit is called the **lumen**. We usually denote the luminous flux in a beam by the letter $F$, and $k_m$ has the value 683 lumens per watt. So for any beam of light the luminous flux is:

$$F = 683 \int_0^\infty P_\lambda V_\lambda \, d\lambda \text{ lumens.} \qquad (1)$$

"Luminous flux" can be thought of as the measure of the amount of visible light in a beam as light travels from place to place—a measure of "light in transit" if you like. An ordinary 100 watt pearl bulb produces a light output of about 1250 lumens; an 80 watt fluorescent lamp of one of the more efficient types will give about 5600 lumens, and a 90 watt low-pressure sodium street lamp (the very yellow sort) will push out over 12 000 lumens.

Notice here the importance of the "spectral power distribution" of the light source in producing light. By "spectral power distribution" (or SPD for short) we mean the relative amount of power emitted at all the different wavelengths of the spectrum. Figure 1.3 shows the spectral power distribution for a projector lamp. In the example just given, the pearl bulb can be seen to be relatively inefficient—that is because the SPD has its peak emission at about 1500 nm right outside the visible spectrum. The sodium lamp is highly efficient because almost all of its energy is emitted at one wavelength (589 nm) which almost coincides with the peak of the visibility ($V_\lambda$) curve.

## 1.1.2.2. Illuminance

Although the measurement and concept of luminous flux is basic, it relates to light passing from place to place. However we are usually more interested with the intensity of light arriving at a surface; the lighting engineer may be concerned with the intensity of illumination on an office desk, or the botanist with the intensity of illumination on the leaves of a plant, but it is the **intensity** of illumination that is of interest. By that we mean the amount of light per unit area—the luminous flux falling on unit area of the surface concerned. This quantity is referred to specifically as the "illuminance" of the surface at the point concerned.

In practice it is almost impossible to obtain uniform illuminance over an area of any size and when illuminance values are quoted they refer to a specific point unless otherwise stated. (Lighting engineers have specific procedures for averaging illuminance values over large areas (CIBSE, 1984)). We define illuminance in the following way.

Suppose that, as in Fig. 1.6, we have a small cone of light coming from a

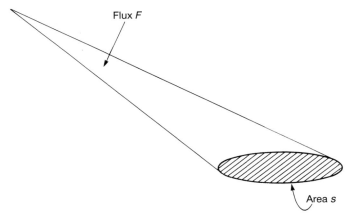

Flux *F*

Area *s*

**Fig. 1.6** Flux in a narrow cone from a single point falling on a flat surface.

single point and falling on an area of the surface of interest. Then if the luminous flux of light in the cone is *F*, and the area is *s*, the illuminance will be *F/s*. But we have to be concerned with a point. Suppose we make our cone narrower, then the area illuminated will also get smaller but the ratio *F/s* will stay the same. So we define illuminance as the value of *F/s* when the cone is infinitesimally small. Mathematically we can write this as

$$\text{Lim } F/s$$
$$s \to 0$$

or

$$\frac{\mathrm{d}F}{\mathrm{d}s}$$

The unit of illuminance is the lumen per square metre and is called the "lux". In most offices the illuminance values at night, when the lights are on, will be somewhere between 150 and 300 lux. The illuminance on the Earth's surface in the UK on an overcast day will be of the order of 5000 lux. Direct sunlight may reach 50 000 or 100 000 lux.

It is important to consider how the illuminance on a surface will change if it is tilted. Suppose, as in Fig. 1.7, that we have light from a single source and that it is falling square on a surface. If we again think of flux *F* in a very narrow cone falling on area *s*, then the illuminance *E* will be:

$$E = F/s \tag{2}$$

What happens if we now tilt the surface through an angle $\theta$ as in Fig. 1.8? The flux from the source will not change but it will be spread over a larger

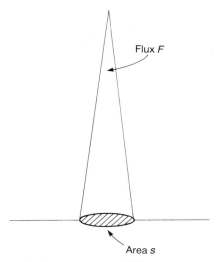

**Fig. 1.7**   Flux in a narrow cone falling square on a surface.

area. In fact the area is increased by the inverse of the cosine of the angle, i.e. we have to replace $s$ by $s/\cos\theta$. So the illuminance is now

$$E = F/(s/\cos\theta),$$

i.e.                                   $$E = F\cos\theta/s \tag{3}$$

and if we call the original value $E_0$ then the illuminance on the tilted surface will be

$$E = E_0\cos\theta. \tag{4}$$

This means that for a given light source, the illuminance on a surface at a given distance depends on the angle it presents. If it is turned away from the square position by $45°$ the illuminance will fall by roughly 30% and if it is turned away by $90°$ the illuminance will fall to zero because the light is now spread out over an infinite area.

This may sound pretty obvious, but it is worth remembering that say, different parts of the same leaf of a plant may receive very different illuminance. Likewise if people lie in brilliant sunlight in the hope of acquiring a tan, the "cosine effect" will usually cause different parts of their body surface to get sunburnt to very different degrees.

In the case where you have light coming from more than one light source, the effect on the illuminance of a surface which is tilted becomes much more complicated. Lighting engineers use special measures such as "spherical

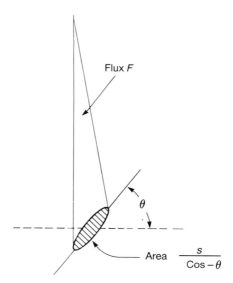

Flux *F*

*θ*

Area $\dfrac{s}{\mathrm{Cos}-\theta}$

**Fig. 1.8**    The effect of tilting the surface in Fig. 1.7.

illuminance" or "cylindrical illuminance" in such cases, but these are beyond our present scope (CIBSE, 1984).


### 1.1.2.3.  Luminous intensity

Having considered light in transit from place to place, and light arriving on a surface, we now have to consider light leaving a source—we have to have some measure of the intensity of a source. Most light sources emit light in almost all directions, but the intensity in different directions is very different. Even the familiar domestic pearl light bulb has markedly different intensities in different directions. So when we specify the intensity we have to do so for a particular specified direction.

Because sources emit light in all directions we have to work in three dimensions, and we have to use "solid angles". Many readers such as physicists and mathematicians will be familiar with solid angles, but for the benefit of those who are not, a digression on solid angles follows.

We are all familiar with ordinary angles, and the way that we measure them in degrees. Look at Fig. 1.9. Imagine that we pivot the line AB at A, and move it round to AC—suppose it makes an angle of 53° to AB. What we mean is that we have rotated it through 53/360 of the angle if we went right round and came back to AB again. It is a purely artificial method of

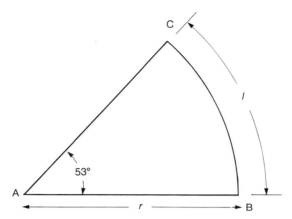

**Fig. 1.9** An angle measured in degrees and radians.

defining an angle, invented as a handy way of doing it by Sir Isaac Newton.

Another way of doing it is that used by mathematicians. As the line AB moves round, the point B traces out the arc BC—suppose this is of length $l$. Then if the radius AB is of length $r$, we can measure the angle BAC in terms of the ratio $l/r$. If AB goes right round once and comes back to the starting position, since the circumference of the circle is $2\pi r$, then the angle measured in this way will be $2\pi r/r$, that is $2\pi$; a right angle would be $\pi/2$, and so on. This technique is referred to as "circular measure" and the unit is called the "radian". Obviously $2\pi$ radians are equivalent to $360°$ and one radian is approximately $57°$.

Now let's think about our light sources, and our original slide projector. That emits light in a very narrow cone but it has both height and width—we could not specify its size with a single figure if we use either degrees or radians. Consider for a moment the type of floodlight used in floodlighting buildings. That will emit light into a very wide cone, which almost certainly has very different angular height and width. So we have to have some other measure; and we use the concept of "solid angles"—angles in three dimensions.

Imagine that our floodlight has a sharp-edged beam and that it is placed at the centre of an imaginary sphere, of radius $r$. Then the light going out from it will fall on a definite area of that sphere, and suppose that area is $A$, as in Fig. 1.10. Then we define the solid angle of its beam as the ratio $A/r^2$—the area divided by the square of the radius. This is exactly the same as circular measure except that we are now working in three dimensions. The units are called steradians, (usually abbreviated to sr) and since the

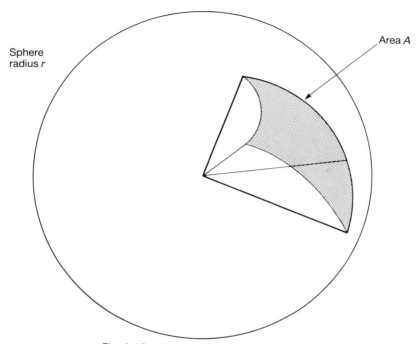

**Fig. 1.10** The measurement of solid angle.

whole surface area of a sphere is $4\pi r^2$, the whole solid angle round a point constitutes $4\pi$ steradians.

We can use solid angle to define the intensity of a light source in a given direction in this way. Suppose that we have a light source (and it may well not be just a lamp, but a complete device like a car headlamp or a lighthouse) and that we consider the light emanating from it in a very narrow cone in the direction of interest. Suppose too that the solid angle of this cone is $\omega$, and the flux contained within it is $F$. Then the "luminous intensity" is defined as the ratio $F/\omega$—the "flux per steradian" in that direction. In mathematical terms the luminous intensity is defined as

$$I = \underset{\omega \to 0}{\text{Lim}}\ F/\omega$$

or

$$I = \mathrm{d}F/\mathrm{d}\omega. \tag{5}$$

The unit of luminous intensity is the "candela" (cd for short). In very old books you may find reference to "candle power" but this term is completely obsolete. The luminous intensity of a 100 watt bulb in its brightest direction will be of the order of 200 candela or so. A lighthouse may have a peak luminous intensity of hundreds of thousands of candela.


### 1.1.2.4.  Luminance

We also need a measure of the brightness of surfaces. Note that this is not the same as the illuminance; "illuminance" refers to light falling on a surface, but here we are concerned with light leaving surfaces—which may include lamps. Think for a moment of a 100-watt clear bulb and a 40-watt fluorescent lamp. They may have roughly the same total flux output and, in some directions, similar luminous intensities—but they are very different to look at. You can look at the fluorescent tube quite comfortably but the incandescent filament of the clear bulb is literally blinding. The vital difference is the light leaving per unit area of surface.

We define the "luminance" of a surface as "the luminous intensity per unit area". That implies the luminance in a given direction; and many "surfaces", e.g. a cinema screen, will have very different luminances in different directions. The unit is the "candela per square metre", usually written as $cd.m^{-2}$.

A typical light-coloured wall in a well-lit office will have a luminance between say 50 and 100 $cd.m^{-2}$. A fluorescent tube may have a luminance of 2500 $cd.m^{-2}$. A compact source xenon arc lamp may have a luminance of 100 000 $cd.m^{-2}$ or so. The luminance of light sources is of importance in the design of optical and spectroscopic instruments, because in many cases it is ultimately the source luminance which limits the amount of light that can be got through the optical system.


### 1.1.2.5.  Practical light measurements

The measurement of light by eye went out (thankfully) about 50 years ago. Modern instruments enable both illuminance and luminance to be measured instantly with sufficient accuracy for most purposes. However, one word of caution is necessary.

Such photometers use photoelectric devices to effect the actual measurement. Bearing in mind that "light" is in fact radiant energy weighted according to its vision-producing effect, it is necessary for the spectral sensitivity of the device to exactly match the spectral luminous efficiency curve. This means that a carefully designed colour filter has to be used to modify the spectral sensitivity function of whatever device (usually referred to as a "detector") is used. It is quite difficult to make filters that will give a precise match. Small errors in this "sensitivity matching" may not matter when white light is used, but can become serious if monochromatic light is being measured, because you then rely on the correction being perfect for that particular wavelength. The instrument maker should be able to provide correction factors for this particular situation.

### 1.1.2.6.  Current and obsolete units

All of the foregoing material is based on the 1987 version of the International Lighting Vocabulary (CIE, 1987) which is published jointly by the Commission Internationale de L'Éclairage and the International Electrotechnical Commission, and is strictly in terms of SI units. However, in many countries SI units are not used—Australia and America for example—with the result that earlier units are still in use; and even in "metric" countries quite a number of pre-SI units are still in use. Amongst the more picturesque of the latter is the "nit" which is in fact the SI unit of luminance, the candela per square metre. Two need specific mention: (i) A "foot-candle" (USA) is a unit of illuminance which is the same as the pre-SI British unit of "one lumen per square foot". It is equivalent to 10.76 lux. (ii) A "Foot-Lambert" or "Ft-L" (USA) is a unit of luminance. One Foot-Lambert is the same as 3.426 $cd.m^{-2}$.

### 1.1.3.  Radiation and radiation measurements

When we speak of "light" we refer to "radiant energy weighted according to its vision-producing effect", but when we are concerned with "radiation" we mean total radiant power. We are often concerned with a wide variety of wavelengths and have to work in terms of their integrated effect but there is no complication of a weighting factor. Take sunlight for example, which extends well into the ultraviolet and infra-red parts of the spectrum. To quantify "radiant flux" we simply have to integrate all the parts of the spectral power distribution—just as was done with luminous flux—and that is all. All the units are in direct SI terms and do not have generic names.

The quantities involved are exactly analogous to those used in light measurement, as described below.

### 1.1.3.1.  Radiant flux

Radiant power travelling from place to place is referred to as "radiant flux". If the radiation in a defined beam contains power per unit wavelength interval of $P_\lambda$ then the "radiant flux" is:

$$P = \Sigma\, P_\lambda\, \delta\lambda \qquad\qquad (6)$$

or

$$P = \int_0^\infty P_\lambda\, d\lambda. \qquad\qquad (7)$$

Theoretically we should have to sum up the power in all wavebands from zero to infinity. For photobiological purposes we can usually neglect any wavelengths shorter than 200 nm because the oxygen of the atmosphere absorbs them. It is dubious whether wavelengths longer say than 10 $\mu$m have any photobiological effects, so in practice we can neglect anything of longer wavelength. The unit of radiant flux is the ordinary unit of power, the watt.

### 1.1.3.2.  Irradiance

Irradiance is analogous to illuminance; it is a measure of the intensity of radiant power falling on a surface at a point, i.e. the amount of radiant flux falling on unit area of a surface. The unit is the "watt per square metre", usually written as $W.m^{-2}$.

If a surface receives radiation from a single point the cosine effect comes into play. For example the irradiance on a surface normal to bright summer sunlight in our latitudes may rise to around 1300 $W.m^{-2}$, but on the horizontal surface of the ground it will perhaps be about 900 $W.m^{-2}$.

### 1.1.3.3.  Radiant intensity

Radiant intensity corresponds to luminous intensity. It is a measure of the radiant power leaving a source per unit solid angle. The unit is the "watt per steradian", usually written as $W.sr^{-1}$.

Most sources of radiation, like light sources, emit with very different radiant intensities in different directions, and it is therefore necessary to specify the direction to which any value of radiant intensity relates.

In the case of lasers, radiant intensity is not normally used. That is because the beam of radiation from a laser effectively does not diverge at all; even at a distance of a kilometre it may be only a few centimetres wide. It would be meaningless to work in terms of steradians. The power output of lasers is specified in terms of the total power emitted (i.e. the radiant flux) in watts. If it is a pulsed laser the output may be expressed in units of energy per flash, i.e. in joules.


### 1.1.3.4. Radiance

Radiance corresponds to luminance, and may be thought of as the radiant power per steradian leaving unit area of a surface. Consequently it needs to be specified in direction, and the point from which it is leaving also must be specified. The unit is the "watt per steradian per square metre", usually written as $W.sr^{-1}.m^{-2}$.

It should be noted that even at room temperatures all surfaces radiate a certain amount of longwave infra-red radiation which contributes to the radiance. For photobiological purposes this is usually of no significance but it should not be overlooked.


### 1.1.3.5. Radiation measurement

Instruments are available for measuring irradiance, radiance and in some circumstances radiant power. However there is one important point about their use which must be mentioned.

Just as a light-measuring instrument must have a precisely defined spectral sensitivity curve, so must radiometric instruments; they must be equally sensitive to radiation of **all** wavelengths. In practice that means they must operate over the wavelength range from 200 nm up to 100 $\mu$m, or so. Now there is no photoelectric device that will work at wavelengths of 100 $\mu$m or anywhere near it; instruments working up to that wavelength have to rely on the heating effect of the radiation—they are called "thermal" detectors and are far from easy to use. For photobiological purposes we are concerned with shorter wavelengths, say up to 5 or 10 $\mu$m, and here in some circumstances some photoelectric detectors can be used. They are combined with colour filters which are supposed to produce a "flat", i.e. uniform, spectral response.

Unfortunately the nature of photoelectric detectors is such that it is difficult to get a flat response at wavelengths above 2 $\mu$m or so. None the less, many of these instruments are described by their manufacturers as "radiometers". If an attempt is made to use any instrument in this category with light sources which produce radiation of wavelength longer than about 1.5 $\mu$m, the spectral sensitivity curve of the instrument (which the manufacturer normally provides) should be checked to see that a meaningful result can be obtained.

### 1.1.4. Photon quantities

I suppose (and hope) that every reader of this book will know something about the "quantum theory" of radiation, developed by Einstein, Planck and others around 1900. Their calculations showed, and many simple experiments demonstrate, that when energy is radiated from one point to another there is not a continuous stream of waves; the radiant energy is made up of an enormous number of small packets of energy, travelling independently. Einstein called these "wave packets" but we usually now refer to them as "quanta" or "photons". For many photobiological purposes it is much more sensible to work in terms of the number of photons per second falling on a surface than in watts. For that reason a third series of quantities has recently been introduced called "photon quantities" (CIE, 1987). Photon quantities are exactly the same as "radiation" quantities except that instead of working in watts, we work in terms of number of photons per second. If for example we take "irradiance" which is a measure of the radiant power falling on unit area of a surface, the corresponding photon quantity will be "photon irradiance" which is the number of photons per second falling on unit area of the surface. Thus "photon flux" corresponds to "radiant flux", "photon intensity" to "radiant intensity", and "photon radiance" to "radiance".

If the radiation concerned is monochromatic or nearly so, conversion from "radiation" to "photon" quantities is straightforward. The energy of a single quantum is given by $W$ in this equation

$$W = h\nu \tag{8}$$

where $h$ is Planck's constant, and $\nu$ is the frequency of the radiation. Since the velocity of any wave must be the product of the frequency and the wavelength, if $c$ is the velocity of light and $\lambda$ is the wavelength then

$$c = \nu\lambda$$

whence

$$W = hc/\lambda. \tag{9}$$

Now $h$ (Planck's constant) has the value $6.626 \times 10^{-34}$ joule seconds and $c$ has the value $2.998 \times 10^8$ m.s$^{-1}$. So if we work out as an example the energy of a single photon of yellow-green visible radiation of wavelength 550 nm it comes to:

$$W = \frac{6.626 \times 10^{-34} \times 2.998 \times 10^8}{550 \times 10^{-9}} \text{ joules}$$

$$= 3.61 \times 10^{-19} \text{ joules}$$

which is not much! If radiation of many wavelengths is involved we have to do this calculation for several wavelengths and sum up the total (CIE, 1987).

### 1.1.5. Other photobiological measurements

We began by talking about light—by which we meant radiant power weighted according to its visual effects. Often in photobiology we have to be concerned with the non-visual effects of radiant energy, such as the induction of erythema or photokeratitis. In many such effects the relative effects of radiation of different wavelengths are known, and in fact a "spectral sensitivity curve" for the effect can be compiled. Such a curve is often referred to as the "action spectrum" for the particular effect.

If such a curve is known, then a series of measures and units for the effect corresponding to those for visible light can be established; the processes are exactly the same except that the $V_\lambda$ function is replaced with the appropriate action spectrum.

### 1.1.6. The spectral power distribution of light sources

In all work in this field it is essential to have a proper knowledge of the SPD of the light sources involved. Light sources divide naturally into various categories.

#### 1.1.6.1. Incandescent sources

Planck showed that in all cases where light is produced by incandescence, the SPD would be predictable and determined by the temperature on the absolute scale, provided that the actual surface of the source would emit all wavelengths equally. Most incandescent sources approximate to that condition; they include such items as radiant electric fires, candle and oil lamp flames, filaments of tungsten lamps and so on. The SPD is given by

$$P_\lambda = C_1/\lambda^5 \left[ (\exp(C_2/\lambda T) - 1 \right] \tag{10}$$

where $\lambda$ is the wavelength, $C_1$ and $C_2$ are constants and $T$ is the absolute temperature. Some typical distributions are shown in Fig. 1.11, (Pivovonsky and Nagel, 1961) and some useful figures for sub-standard tungsten lamps are given by Jones (1970).

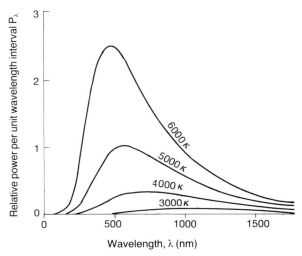

**Fig. 1.11**  Spectral power distribution curves for Planckian ("Black Body") radiators.

### 1.1.6.2. Discharge lamps

Lamps which operate by means of an electrical discharge through a gas usually produce a mixture of a line spectrum and a continuous spectrum. Such lamps include the familiar fluorescent tube. Sometimes so many spectral lines are produced it is difficult to distinguish between line and continuum—xenon arcs being the prime example. Lamp manufacturers will usually supply information about the SPDs of their products gratis—it is usually given in rudimentary form in their catalogues. Since the $P_\lambda$ figure in SPD tables represents power per unit waveband it is necessary to adopt a convention to enable the figures for the lines compatible with that of the continuum. One such is to assume that the power in a single spectral line is spread out over a band 10 nm wide—but other conventions are often used.

### 1.1.6.3. Sunlight, skylight and daylight

When we speak of "sunlight" we normally mean radiation falling on the Earth's surface direct from the Sun. The term "skylight" is reserved for light reaching the Earth's surface from the blue sky or from clouds, and "daylight" is used to mean the integrated effect of sunlight and skylight. The spectral power distribution of daylight varies with several meteorological factors, but principally depends on the proportion of sunlight and skylight received on a given surface. The subject is a complicated one and is fully dealt with in an excellent book by Henderson (1977) and by Björn in this volume (see Chapter 8).

Figures for the relative SPD of various phases of daylight were published by the CIE in the late 1960s but remain unchanged (CIE, 1986), and are widely accepted as far as visible wavelengths are concerned. It is now known that there are wide variations in the ultraviolet content of daylight, especially with altitude (Kok, 1972) and the CIE figures may be too low in the ultraviolet region (Tarrant, 1975).

### 1.1.6.4.   Colour temperature

The SPD of a light source is sometimes referred to by its "colour temperature". The idea is, with incandescent sources, that this shall be the temperature of a source which obeys Planck's law whose SPD most closely resembles that of the source in question. Most incandescent sources show slight departures from Planck's law with the result that the "colour temperature" is slightly different from the actual temperature.

Unfortunately the term "colour temperature" is frequently applied —wrongly—to light sources which are not incandescent ones, e.g. fluorescent lamps, as a crude way of specifying their colour. The term should only be applied to incandescent sources and whilst it is much used in photographic circles, it is of very limited value in photobiological work, because it specifies nothing about the SPD outside the visible region.

## References

Chartered Institution of Building Services Engineers (1984). Code for Interior Lighting Practice. CIBSE, 222 Balham High Road, London SW12 9BS.

"Colorimetry". (1986). 2nd Edition, CIE Publication No. 15.2, Bureau Central de la CIE, Vienna.

Commission Internationale de l'Eclairage (1987). "International Lighting Vocabulary", 4th Edition. CIE Publication No. 17.4, Bureau Central de la CIE, Vienna.

Henderson, S. T. (1977). "Daylight and its Spectrum", 2nd Edition. Adam Hilger, Bristol pp. 1967–1976.

Henderson, S. T. & Marsden, A. M. (1972). "Lamps and Lighting", 2nd Edition, p. 30. Edward Arnold, London.

Jones, O. C. (1970). Standard spectral power distributions. *J. Phys. D: Appl. Phys.* **3**, 1967–1976.

Kok, C. J. (1972). Spectral irradiance of daylight for air mass 2. *J. Phys. D: Appl. Phys.* **5**, 185–188.

Pivovonsky, M. and Nagel, M. (1961). "Tables of Black-body Radiation Functions", Macmillan (USA).
Tarrant, A. W. S. (1975). Further studies of the spectral power distribution of daylight in the ultraviolet region. Compte Rendus 18th Session CIE, London, pp. 384–392. CIE Publication No. 36.

## Note added in proof

CIE publications may be obtained in the UK from Thorn Lighting Ltd., Enfield, Middlesex. (Contact Mr. Dangerfield.)

# 2
# Optical Radiation Detectors

## A. D. WILSON

*Applied Physics Group*
*Pilkington Optronics*
*Barr & Stroud Ltd*
*Caxton Street*
*Anniesland*
*Glasgow G13 1HZ, UK*

## Nomenclature

| | |
|---|---|
| $A$ | area |
| $B_s$ | signal bandwidth |
| $C$ | capacitance |
| $C_a$ | input capacitance of amplifier |
| $C_d$ | capacitance of detector |
| $c$ | heat capacity |
| $c$ | velocity of light |
| $D$ | detectivity |
| $D^*$ | normalized detectivity |
| $E_g$ | band gap of semiconductor |
| $E_a$ | electron affinity |
| $e$ | electronic charge |
| $f$ | frequency |
| $\Delta f$ | noise bandwidth |
| $G$ | gain |
| $g$ | thermal conductance |
| $h$ | Planck's constant |
| $i$ | current |
| $i_{ad}$ | anode dark current |
| $i_{cd}$ | cathode dark current |
| $i_{ct}$ | thermionic component of $i_{cd}$ |
| $i_{cx}$ | extrinsic component of $i_{cd}$ |
| $i_n$ | noise current |
| $i_0$ | reverse saturation current of a photodiode |
| $i_{DAN}$ | detector amplifier noise current |
| $i_{JN}$ | Johnson noise current |
| $i_{GR}$ | generation–recombination noise current |

RADIATION MEASUREMENT IN PHOTOBIOLOGY
ISBN 0–12–215840–7

| $i_{SN}$ | shot noise current |
| $K$ | electron multiplier noise factor |
| $k$ | Boltzman's constant |
| $N$ | number of thermocouple junctions in a thermopile |
| NEP | noise equivalent power |
| $n$ | diode ideality factor |
| $P$ | radiant power |
| $P_s$ | polarization |
| $p$ | pyroelectric coefficient |
| $\mathscr{R}$ | responsivity |
| $R$ | resistance |
| $R_f$ | feedback resistance |
| $R_L$ | load resistance |
| $R_{d0}$ | zero-bias slope resistance of a photodiode |
| $r$ | temperature coefficient of resistance |
| $S$ | Seebeck coefficient |
| $T$ | temperature |
| $\Delta T$ | temperature rise |
| $V$ | voltage |
| $\Delta V$ | voltage increment |
| $V_{JN}$ | Johnson noise voltage |
| $V_n$ | noise voltage of amplifier |
| $Z_d$ | detector impedance |
| $\alpha$ | absorptance |
| $\alpha_R$ | Richardson constant |
| $\varepsilon$ | critical energy barrier/gap in a detector material |
| $h\nu$ | photon energy |
| $\phi$ | work function |
| $\Phi$ | barrier height |
| $\tau$ | rise time |
| $\tau_E$ | electrical time constant |
| $\tau_e$ | electron transit time |
| $\tau_T$ | thermal time constant |
| $\lambda$ | wavelength |
| $\lambda_{co}$ | cut-off wavelength |
| $\omega$ | angular frequency |
| $\omega_a$ | gain-bandwidth product of amplifier |
| $\eta$ | quantum efficiency |

## 2.1.  Introduction

An optical radiation detector consists of an element which absorbs the
radiation and a means of measuring the resulting change in some property
of the element. There are two main classes of detector (thermal and photon)

which utilize radically different mechanisms for interaction of the radiation with the element.

In thermal detectors, the absorption of radiation increases the temperature of the element and this change can be detected by a variety of means, e.g. using a thermocouple, a direct change in the electrical resistance of the element, a change in surface charge due to the pyroelectric effect, a change in gas pressure etc. Since it is possible to make an element which absorbs nearly equally at all optical wavelengths, thermal detectors can have a broad spectral response with a near-uniform sensitivity.

Photon detectors utilize quantum effects. There is no effect unless the energy of the photon exceeds some critical value, $\varepsilon$, and thus such detectors have a restricted spectral response with a responsivity cut-off at an upper wavelength, $\lambda_{co}(\mu m)$, equal to $1.241/\varepsilon(eV)$. Examples include photomultipliers, photodiodes and photoconductors. In each case, charge carriers are generated upon photon absorption and can be detected in external electronic circuitry.

The wide range of available photodetectors presents the prospective user with a complex choice. Detector parameters which need to be considered include the radiant sensitivity, spectral response, speed of response, noise performance (and hence signal to noise ratio), linearity at high signal levels, size, cost and the optimum mode of detector operation including external electronics. It is the purpose of this chapter to review the various types of detectors and when and how to use them. After a brief description of common detector performance parameters and noise sources, there are sections dealing with each of the types of thermal and photon detectors likely to be encountered in photobiophysical studies. Finally, available detector formats are described and several specific applications are examined.

## 2.2. Detector Performance Parameters

Before discussing specific types of detectors it is necessary to define a number of commonly used detector performance parameters (see Dereniak and Crowe, 1984).

The responsivity, $\mathscr{R}$, is the ratio of the electrical output to the radiant input: the units are $A.W^{-1}$ or $V.W^{-1}$. The noise equivalent power, NEP, is that required to produce a signal equal to the root mean square (rms) noise signal. If $i_n$ is the rms noise current (in A) then the NEP $= i_n/\mathscr{R}$ and has units of W. Often the NEP is quoted in units of $W.Hz^{-1/2}$: to convert to a true NEP simply multiply by the square root of the noise bandwidth, $\Delta f^{1/2}$. This latter should not be confused with the measurement frequency. For example, if radiation is chopped at a frequency of 1 kHz and the detector output is filtered to give a bandpass between 995 and 1005 Hz then $\Delta f$ is 10 Hz. Nor should it be confused with the signal bandwidth, $B_s$, of a broadband measurement system: for a single pole system, $\Delta f \approx (\pi/2)B_s$ (Horowitz and Hill, 1980). The detectivity, $D$, is the reciprocal of the NEP. The normalized detectivity, $D^*$, is defined as

$$D^* = DA^{1/2}\Delta f^{1/2} = A^{1/2}\Delta f^{1/2}/\text{NEP}. \tag{1}$$

where $A$ is the detector area: the units are $\text{cm.Hz}^{-1/2}.\text{W}$. $D^*$ can be used to assess the performance of different detector types independent of device area and noise bandwidth. For this to be valid $i_n$ must be proportional to $A^{1/2}$ and $\Delta f^{1/2}$. Since this is not always the case and since real detectors often have widely differing active areas (cf. photomultipliers and photo-diodes), the NEP will often be the more useful guide to detector selection.

Since detector noise is clearly a critical parameter it is useful to examine the more common noise sources. In what follows, $e$ is the electronic charge $(1.6 \times 10^{-19}$ coulomb), $T$ is the absolute temperature and the noise currents (voltages) are given as rms values (peak to peak values are $6 \times$ rms values).

Shot noise in a current, $i$, is due to the quantized nature of charge generation and is given by

$$i_{SN} = (2ei\Delta f)^{1/2}. \tag{2}$$

It applies when generation involves emission over a potential barrier: either carriers of a single polarity are involved, as in photoemissive detectors, or if electron-hole pairs are produced they do not recombine, as in ideal photodiodes. If, as in photoconductors, the electron-hole pairs can recombine, the recombination leads to increased noise: this generation–recombination noise current is given by

$$i_{GR} = (4ei\Delta f)^{1/2}. \tag{3}$$

The current $i$ may be the device dark current, the signal photocurrent or the background photocurrent. This noise source only applies to photon detectors.

Johnson noise, due to thermal fluctuations in a resistor, is given by

$$i_{JN} = (4kT\Delta f/R)^{1/2}: \quad V_{JN} = (4kT\Delta fR)^{1/2}. \tag{4}$$

where $R$ is the resistance of the detector element. This source is common to both photon and thermal detectors. There is also a contribution from any external measuring resistors.

Often these sources are the only ones referred to in manufacturers' data sheets. However in many cases, the role of the amplifier must be considered. The interaction of the detector impedance, $Z_d$, with the voltage noise of the amplifier, $V_n$, gives rise to an additional detector–amplifier noise term,

$$i_{DAN} = V_n/Z_d. \tag{5}$$

Since both $V_n$ and $Z_d$ can be frequency dependent, the value of $i_{DAN}$ may depend on the measurement frequency and noise bandwidth: see Hamstra and Wendland (1972) and Wilson and Lyall (1986a).

Other noise sources which may cause problems are $1/f$ noise in low frequency measurements and line (mains)-borne electrical interference: the latter can be minimized by appropriate decoupling of amplifiers from power supplies.

## 2.3. Thermal Detectors

### 2.3.1. Basics

As has already been noted, thermal detectors are inherently broadband. Many are able to detect from the UV through to millimetre wavelengths when used without windows. The latter tend to restrict the spectral response. Almost all thermal detectors operate at room temperature.

The basic thermal detector structure is an absorbing element of heat capacity, $c$, and optical absorptance, $\alpha$, plus a temperature sensor connected to a temperature sink, the connection having a thermal conductance, $g$. The temperature rise of the element relative to the sink, $\Delta T$, depends on the balance between the rates of heating by radiation absorption and of heat loss through the conductance. This process is time dependent and thus $\Delta T$ depends on the modulation frequency ($\omega = 2\pi f$) such that

$$\Delta T = (P\alpha/g)(1 + \omega^2\tau_T^2)^{-1/2} \tag{6}$$

where $P$ is the incident radiant power and $\tau_T = c/g$ is called the thermal time constant. The frequency dependence of $\Delta T$ is controlled by $\tau_T$, the reciprocal of which is called the thermal corner frequency. At very low frequencies ($\omega\tau_T \ll 1$), the temperature rise is independent of frequency, i.e. $\Delta T = P\alpha/g$. At higher frequencies ($\omega\tau_T \gg 1$) the temperature rise is inversely proportional to frequency, i.e. $\Delta T = (P\alpha/c)\omega^{-1}$ and thus the detector's ability to faithfully respond to a chopped input is limited.

The poor temporal response of thermal detectors does not prevent their use in measuring the energy of laser pulses with a responsivity independent of pulse length since in this mode the detector acts as an integrator.

A maximized responsivity is obtained by using a highly absorbing element ($\alpha \to 1$) and a low value of $g$. Good high frequency response needs a low heat capacity implying a small element.

In the absence of radiation the element should be at the temperature of the sink. However, there are always micro-fluctuations in the element temperature which leads to temperature fluctuation noise (Smith *et al.*, 1957). Other noise sources dominate the performance of most thermal detectors.

The types of thermal detectors differ in the means used for measuring the temperature rise of the element. The two major types of detector are the thermopile and the pyroelectric, both of which can be of very rugged construction. Of lesser importance are the bolometer, Golay and photo-acoustic detectors.

### 2.3.2. The thermopile

The temperature rise, $\Delta T$, can be sensed using a thermocouple, a junction formed from dissimilar metals across which a small voltage is generated upon heating. Normally a number, $N$, of junctions are connected in series to give a larger output than can be obtained from a single junction. This is known as a thermopile. Its output voltage, $\Delta V$, is given by $\Delta V = NS\Delta T$ where $S$ is the Seebeck coefficient of the junction. In most modern thermopiles the junctions are formed from evaporated films of Bi and Sb since these materials give the largest $S$ values: $100\,\mu\text{V.K}^{-1}$. The voltage responsivity, $\mathcal{R}_v$, of a thermopile is simply $\Delta V/P$ and thus has the same frequency dependence as was discussed previously for $\Delta T$. Thermopiles can be designed to have response times from sub-microsecond to many seconds: typical instrumentation grade devices have values around 50 milliseconds giving corner frequencies of around 3 Hz. At frequencies below this value, $\mathcal{R}_v$ tends to a value of about $20\ \text{V.W}^{-1}$. Using such a thermopile at frequencies above 100 Hz is not advisable.

To obtain optimum performance the thermopile should be impedance matched to a suitable amplifier. Wire junction thermopiles have low resistances (50 $\Omega$) and must be coupled to the amplifier using a transformer. Thin film devices have relatively high resistances (5 k$\Omega$) and thus can be coupled directly to high quality bipolar operational amplifiers. Since the thermopile output is mV or less the main requirements on the amplifier are a low input offset voltage and a low temperature drift in that parameter coupled with a low noise voltage, especially at frequencies less than 20 Hz. The OP-27 and 7650 series of op-amps are suitable. Alternatively, the small signal voltages can be measured using a galvanometer (Smith *et al.*, 1957).

The main noise source is Johnson noise in the thermopile resistance. Typically, the NEP will be approx. $1\ \text{nW.Hz}^{-1/2}$. Linear behaviour over 5 decades of input power can be achieved.

### 2.3.3. The pyroelectric detector

Here, the temperature sensor is a slab of ferroelectric material placed between two electrodes. The materials most commonly used are triglycine sulphate, lead zirconate–titanate and lithium tantalate. A black coating is often used to give a broad spectral response. The slab exhibits a polarization, $P_s$, formed during a poling process (Rose, 1982). With the material above its Curie temperature an electric field is applied in a direction normal to the electrodes to maximize $P_s$. The material is cooled to room temperature whence the field is removed. When this poled detector slab is heated by radiation the induced polarization and consequently the charge on the electrodes varies. The temperature of the slab must never exceed the Curie temperature otherwise the poling process is undone and the detector ceases to function. The change in the charge on the electrodes causes a compensating current flow in external circuitry. An output is only produced when the temperature is changing and thus the pyroelectric detector only responds to modulated or pulsed radiation. In this, it is different to other thermal detectors, such as thermopiles and bolometers.

The current responsivity, $\mathscr{R}_i$, of a pyroelectric is given by

$$\mathscr{R}_i = \omega A p \Delta T / P \tag{7}$$

where $A$ is the detector area and $p$ is the pyroelectric coefficient which is $\partial P_s / \partial T$ and has units of coulomb.m$^{-2}$.K$^{-1}$. A typical value for $\mathscr{R}_i$ is $1~\mu$A.W$^{-1}$. This value is much lower than found with photon detectors such as silicon photodiodes where the value is 0.5 A/W (see Section 2.4.3). Since the pyroelectric coefficient varies with temperature the responsivity is temperature dependent.

It is normal practice to discuss the responsivity of pyroelectrics in terms of the voltage responsivity, $\mathscr{R}_v$. To convert $\mathscr{R}_i$ to $\mathscr{R}_v$, one simply multiplies by the resistance, $R$, appropriate to the mode of operation. In the so-called voltage mode, a load resistor is connected across the electrodes (Fig. 2.1a): an FET source follower is often added as an impedance converter. The value of $R$ is the parallel combination of the detector resistance, $R_d$, and the load resistance, $R_L$. Normally, $R_d \gg R_L$ and thus $R = R_L$. In the so-called current mode (Fig. 2.1b) the detector is connected across the inputs of a JFET op-amp used in the (inverting) transimpedance configuration. In this case, $R = R_f$.

Normally, pyroelectric elements have large values of $\tau_T$. When this is coupled with a high electrical time constant ($\tau_E = C_d R_L$ in voltage mode) the result is a high $\mathscr{R}_v$ value at low frequencies (up to $10^3$ V.W$^{-1}$) with a fall-off inversely proportional to frequency. As $R_L$ is decreased, $\mathscr{R}_v$ decreases and a flat region appears in the $\mathscr{R}_v$–frequency curve (Fig. 2.2).

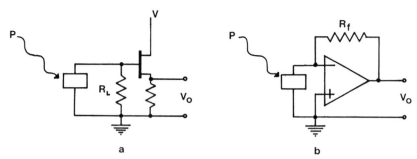

**Fig. 2.1**  Pyroelectric detector operation: (a) voltage mode with an FET follower; (b) current mode using an op-amp. P, radiant power; $R_f$, feedback resistor; $R_L$ load resistor; $V_0$, output voltage.

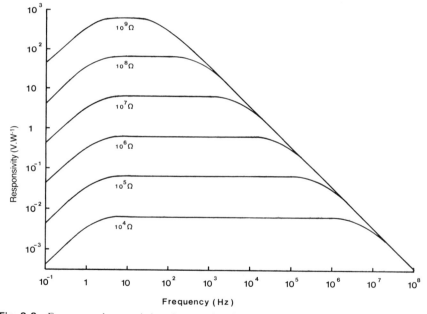

**Fig. 2.2**  Frequency characteristics of a pyroelectric detector used in voltage mode. $R_L$ values are indicated.

A bandwidth of 10 kHz with $\mathscr{R}_v \approx 1$ V.W$^{-1}$ is readily obtained, although care is required to ensure that any black coating does not have a high heat capacity which would reduce the bandwidth (Blevin and Geist, 1974). The pyroelectric detector is unique among thermal detectors in being able to give a frequency independent response. In current mode, a similar shaped $\mathscr{R}_v-$

frequency plot is obtained but the bandwidth can be higher. For this case the high frequency part of Fig. 2.2 is oversimplified since the limiting bandwidth of the amplifier increases the slope of the roll-off.

Johnson noise dominates pyroelectrics at low frequencies: the noise voltage, to a first approximation, is given by

$$V_{JN} = (4kTR\Delta f)^{1/2}(1 + \omega^2 R^2 C^2)^{-1/2} \tag{8}$$

where $C$ is the parallel capacitance of the detector, $C_d$, and the amplifier input, $C_a$. (NB, typically $C_d \approx 30$ pF, $C_a \approx 5$ pF.) At higher frequencies, amplifier noise may dominate. In the Johnson noise regime NEP values of $1-100$ nW.Hz$^{-1/2}$ are readily obtainable. A dynamic range in excess of $10^6$ is possible.

Since all pyroelectric materials are also piezoelectric, problems can be encountered with acoustic pick-up. This can be severe at frequencies near the mechanical resonance frequency of the pyroelectric slab (Glass and Abrams, 1970). This effect can be troublesome when detecting fast laser pulses ($< 1$ $\mu$s) but is less of a problem in lower frequency applications.

### 2.3.4. Other thermal detectors

In the bolometer, the temperature rise is sensed as a change in resistance of the absorbing element which can be a thin metal film, a sintered mixture of metal oxides (the thermistor) or a semiconductor crystal. The voltage output is given by

$$V = irR\Delta T \tag{9}$$

where $i$ is a DC bias current, $R$ is the electrical resistance and $r$ is the temperature coefficient of resistance of the element. Thermistor bolometers work at room temperature, have response times of milliseconds, low frequency $\mathscr{R}_v$ values of 1000 V.W$^{-1}$ and NEPs of 1 nW. Semiconductor bolometers have sub-millisecond response times, slightly higher responsivities and NEPs of 1 pW but have to be cooled to liquid He temperature and thus only find use in specialist applications such as astronomy.

In the Golay cell, or pneumatic detector, the temperature rise is sensed by the expansion of an enclosed gas which presses against a flexible mirror altering its focal length (Golay, 1952). This change is read optically.

Photoacoustic detectors use a microphone to detect the small pressure fluctuations which occur when the heat generated by radiation absorption is coupled into the gas contained in an acoustic resonator (Satheeshkumar and Vallabhan, 1985).

## 2.4. Photon Detectors

### 2.4.1. Basics

Photon detectors utilize quantum effects. Charge carriers are created only when the photon energy exceeds a critical value, $\varepsilon$: $\varepsilon$ is the bandgap of a semiconductor or some barrier height. Consequently, the responsivity of a photon detector extends from an upper cut-off wavelength, $\lambda_{co}$ ($\mu$m), equal to $1.241/\varepsilon$ (eV) to some lower wavelength limit which in practice is dictated by window materials or the detailed physics of the device (Sze, 1969). The

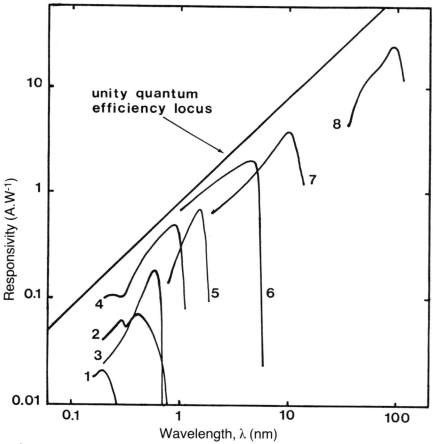

**Fig. 2.3** Current responsivities of photon detectors. Key: 1, CsTe photocathode; 2, S10 photocathode; 3, GaAsP photodiode; 4, Si photodiode; 5, Ge photodiode; 6, InSb photodiode; 7, HgCdTe photoconductor; 8, Ge:Ga photoconductor.

current responsivity, $\mathscr{R}_i$, is given by

$$\mathscr{R}_i = \frac{\eta e \lambda}{hc} \tag{10}$$

where $h$ is Planck's constant ($6.63 \times 10^{-34}$ J.s$^{-1}$), $c$ is the velocity of light ($3 \times 10^8$ m.s$^{-1}$), $e$ is the electronic charge and $\lambda$ is the wavelength. The quantum efficiency, $\eta$, is the number of electrons reaching the electrodes of the detector divided by the number of photons striking the detector surface and is a measure of the efficiency of the carrier generation and collection processes. Rarely does $\eta$ approach the ideal value of unity. This may be because electron-hole pairs generated by the photon absorption event recombine or because the surface reflectance is unlikely to be zero throughout the responsivity range. An internal quantum efficiency is often defined which eliminates the effects of reflectance (Geist and Zalewski, 1979). The current responsivity increases proportionally with wavelength and, for $\eta = 1$, is 0.8 A.W$^{-1}$ at $\lambda = 1$ $\mu$m and 8 A.W$^{-1}$ at $\lambda = 10$ $\mu$m. This trend is shown for several real detectors in Fig. 2.3. Even the poorest photon detector has superior values of $\mathscr{R}_i$ (and $D^*$) to a thermal detector.

Like thermal detectors, many photon detectors can be operated at room temperature. However, cooling becomes necessary when the detector material's critical energy, $\varepsilon$, is small, since thermal generation of carriers may otherwise swamp any photo-excitation effects. As a rule of thumb, if $\varepsilon > 1$ eV ($\lambda_{co} < 1.2$ $\mu$m), room temperature operation is acceptable; if $\varepsilon = 0.4$ eV ($\lambda_{co} \approx 3$ $\mu$m) cooling to 193 K is advisable and if $\varepsilon = 0.12$ eV ($\lambda_{co} \approx 10$ $\mu$m) cooling to 77 K is essential. In critical low-noise applications cooling can be beneficial even when $\varepsilon > 1$ eV. Methods for cooling detectors have been reviewed by Chiari and Morten (1982). Detector manufacturers can often supply cooling systems matched to their own detectors.

### 2.4.2. Photoemissive detectors: the vacuum phototube and the photomultiplier

These detectors have a photocathode from which electrons are emitted on absorption of photons, subject to the constraints of the quantum efficiency. The electrons are collected by an anode and appear as a photocurrent. If the photocathode is a metal then the photon energy, $h\nu$, must overcome the metal's work function: $h\nu > \phi$. Most metals have $\phi > 3$ eV and thus only respond in the ultraviolet ($\lambda < 0.4$ $\mu$m): they also tend to have very low quantum efficiencies. A longer wavelength response can be obtained using semiconductor materials such as $Cs_3Sb$, $BiOAgCs$ and $Na_2KSb$. The photon energy has to overcome both the band gap energy and the electron affinity:

$h\nu > E_g + E_a$. Typically $E_g + E_a$ ranges from 1 eV ($\lambda_{co} \approx 1.2\ \mu$m) to 2 eV ($\lambda_{co} \approx 0.6\ \mu$m) and the peak quantum efficiency can exceed 0.2 for the materials noted above. A third type of photocathode is based on GaAs which has $E_g + E_a \approx 4.3$ eV and thus would be thought to be of limited use. However by applying a thin layer of $Cs_2O$ to the GaAs surface the energy bands are bent such that the vacuum level is below the conduction band in the semiconductor bulk and the electron affinity becomes negative. Photo-emission then occurs when $h\nu > E_g$ which is 1.4 eV for GaAs and thus the device responds out to 0.88 $\mu$m. This is known as a negative electron affinity (NEA) system. A detailed description of the physics of photocathode processes is given by Seib and Aukerman (1973).

Electrons are also emitted from a photocathode kept in the dark and result in a cathode dark current given by

$$i_{cd} = i_{cx} + i_{ct}: i_{ct} = \alpha_R A T^2 \exp(-\Phi/kT) \tag{11}$$

where $i_{cx}$ is the extrinsic component due to cosmic rays, background radioactivity and electrical leakage and $i_{ct}$ is given by the Richardson equation for thermal emission of electrons over an energy barrier, $\Phi$ ($= \phi$ for metals and $E_g + E_a$ for semiconductors): $A$ is the area of the cathode and $\alpha_R$ is the Richardson constant. Cooling the photocathode reduces $i_{ct}$, but, for all but the S1 type photocathodes, below about $-30°$C, $i_{cx}$, which is independent of temperature, dominates and there is no advantage in further cooling. Further decreases in the dark current can only be achieved by reducing the detector area and by choosing a photocathode with an increased value of $\Phi$, thereby restricting the spectral response.

There are three types of photoemissive detector, differing in the way in which the electrons emitted from the photocathode (PC) are collected.

(i) The vacuum phototube. In this simplest device, a voltage is applied between the PC and the anode, which are placed in an evacuated tube (Fig. 2.4). The field ensures that most of the electrons are collected. The gain, defined as the number of electrons collected at the anode to the number emitted by the PC, is effectively unity. The device has a low responsivity ($< 0.05$ A.W$^{-1}$) but is inherently fast (risetime $< 1$ ns) and so is often used for monitoring visible/near infra-red laser pulses.

(ii) The gas-filled phototube. The electrons accelerated in the applied field gain sufficient energy to ionize the gas molecules on impact thereby producing more electrons. Gains of around 10 are feasible.

(iii) The photomultiplier tube. This is the major type of photoemissive detector (Fig. 2.4). The electrons from the PC are accelerated and focused into an electron multiplier, which consists of a series of electrodes, called dynodes, maintained at increasingly positive potentials with respect to the PC. This is achieved through use of a chain of dynode-biasing resistors: the

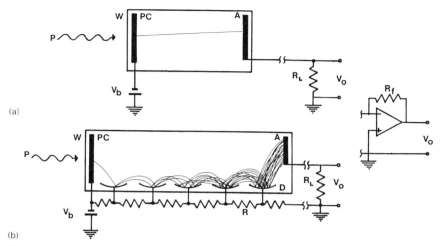

**Fig. 2.4** Photoemissive detectors: (a) vacuum phototube; (b) photomultiplier. A, anode; D, dynode; P, radiant power; PC, photocathode; R, dynode resistor; $R_f$, feedback resistor; $R_L$ load resistor; $V_b$, bias voltage; $V_o$, output voltage; W, window.

voltage divider network. Each time an electron strikes a dynode it causes emission of several secondary electrons which are in turn accelerated to strike the next dynode to give yet more secondary electrons, and so on. Very large gains can be achieved: a typical value is $10^6$. Both the cathode current responsivity and the cathode dark current are multiplied by the gain to give the anode equivalents. Typically, for a gain of $10^6$, the anode current responsivity is $5 \times 10^4$ A.W$^{-1}$ and the anode dark current is in the range 100 pA to 10 nA.

The vacuum phototube (VPT) and the photomultiplier tube (PMT) differ significantly in their responses to fast optical pulses. In the VPT, the photoelectrons travel directly from the cathode to the anode in typically 1 ns. This is the electron transit time, $\tau_e$. In the PMT, $\tau_e$ is increased (20–120 ns) due to the convoluted electron paths through the electron multiplier, i.e. the current pulse is significantly delayed with respect to the light pulse. PMTs also show a spread in $\tau_e$ due to individual electrons following slightly different paths through the multiplier: values of 2–50 ns are found. This can lead to distortion of the pulse shape. Finally, the rise time of the anode photocurrent in the PMT varies from 1 to 20 ns. The temporal performance of a PMT depends critically on the electron multiplier structure. There are four main designs: circular or compact focused; linear focused; box and grid; venetian blind. Linear and compact focused types have the fastest response. Other parameters including gain, high current linearity, spatial uniformity of response (generally, not very good) and immunity from magnetic interference, also depend upon the electron multiplier structure and also on the shape of the photocathode (Hamamatsu, 1979; Thorn-EMI, 1986)

In view of the distortion/delay in pulse response one could ask why PMTs are used much more than VPTs. The answer is that the electron multiplier is a high gain, high bandwidth, virtually noiseless amplifier with which to transform small photocathode currents to more easily measured current values. It would be very difficult to emulate this level of performance using a VPT and an external solid state amplifier.

The ultimately limiting noise source in a PMT is shot noise in the photocathode dark current. The noise current, at the anode, is

$$i_N = KG(2ei_{cd}\Delta f)^{1/2} = K(2ei_{ad}G\Delta f)^{1/2} \tag{12}$$

where $i_{ad}$ is the anode dark current, $G$ is the gain and $K$ is an additional noise factor due to noise in the electron multiplication process: $K$ is usually less than 1.3. For a PMT with a dark current of 1 nA and a gain of $10^6$, $i_N = 20$ pA.Hz$^{-1/2}$ and the NEP is $4 \times 10^{-16}$ W.Hz$^{-1/2}$. Often shot noise in the photocurrent will dominate: simply replace $i_{cd}$ and $i_{ad}$ in Equation (12) by the sum of the respective signal and dark currents. With VPTs, shot noise in the cathode dark current is much less than Johnson noise in the load resistance and/or amplifier noise.

At low light levels the signal current may be less than the dark current. In DC measurements, one has to rely on trimming out the dark current by signal injection with allowance for the temperature dependence of the dark current. If the light can be modulated then capacitive coupling or synchronous detection using a lock-in amplifier will eliminate the dark current and any slow changes due to temperature drift.

As we have already noted, the linearity of the PMT depends on the dynode structure and the shape of the photocathode. In addition, the type of dynode voltage divider network is important. It is usually stipulated that the anode current should never exceed 1/100 of the current in the biasing resistors. This limits the anode current to 100 $\mu$A in DC measurements and approx. 10 mA in measurement of fast pulses. One per cent linearity can be expected over a dynamic range of more than $10^4$. This aspect and others such as drift, fatigue and hysteresis in PMTs are discussed in manufacturers' data sheets (Hamamatsu, 1979; Thorn-EMI, 1986).

Much of the previous discussion relates to the so-called electrometer mode of operation of photomultipliers. This is appropriate for cathode currents of more than $10^{-16}$ A (or 600 electrons per second). At lower levels the technique of photon counting is used (Meade, 1981). Each photoelectron gives rise to a single anode current pulse having a magnitude lying within a very small range of values. Since background pulses have a much wider range of magnitudes, a pulse height discriminator can be used to separate the "photon" pulses from the noise. The technique can measure down to less than 1 photon per second but to do this needs long integration times. Manufacturers produce special photomultipliers for photon-counting applications.

### 2.4.3. Junction photodetectors

Photodiodes, or junction photodetectors, have a depletion region associated with either a p–n junction, formed when n-doped and p-doped material are in contact, or a Schottky barrier, formed when a thin metal film is deposited on a semiconductor surface. The depletion region is so called because it is depleted of mobile carriers. The space charge left behind upon depletion sets up a large electric field across the region (Sze, 1969). It is this field which is the key to photodiode operation.

On absorption of a photon with an energy greater than the band gap of the semiconductor ($h\nu > E_g$) an electron-hole pair is formed. If these carriers are generated within the depletion region they are quickly separated and swept out by the field and current flows in an external circuit. If the carriers are created outwith the depletion region they must diffuse (a slow process compared to depletion field sweep out) into the diffusion region before being collected and thus may be lost by recombination. For maximum responsivity and high-speed operation at all wavelengths below the cut-off value, the width of the depletion region should be as large as possible.

Whether a photon will be absorbed in the depletion layer depends on the absorption coefficient of the semiconductor and the structure of the photodiode. For example in silicon, the absorption depth (the reciprocal of the absorption coefficient) is $< 0.1\ \mu$m for UV photons but increases with wavelength becoming $10\ \mu$m at $\lambda = 0.8\ \mu$m and $> 100\ \mu$m for $\lambda > 1\ \mu$m. Thus the depth of the junction has a critical effect on the spectral response, with shallow junctions favouring a high short wavelength response and deep junctions having only a long wavelength response. A shallow junction with high UV sensitivity is readily achieved with Schottky barrier devices (Wilson and Lyall, 1986b).

The depletion layer width depends on the resistivity of the semiconductor material. Si solar cells are made of low resistivity material and thus have narrow junctions. For this reason and because the low resistivity leads to poor noise performance, they should not be used for accurate radiometry. Conventional pn photodiodes use higher resistivity material and thus have wider depletion regions. By interposing a very high resistivity (near-intrinsic) layer between the p and n layers (the pin photodiode) the depletion region is further extended and these devices are capable of high speed, broadband (0.2–1.1 $\mu$m) operation. A further increase in the depletion width is achieved by reverse biasing the junction.

The best way of using a photodiode can be seen by considering the current–voltage (*IV*) characteristics. The current, *i*, is given by

$$i = i_0 \left[ \exp\left(\frac{eV}{nkT}\right) - 1 \right] - \mathcal{R}_i P. \tag{13}$$

The first term is the dark current: at zero bias there is no dark current: when reverse biased a dark current flows with the value $i_0$, the reverse saturation current, if the applied voltage, $V$, is sufficiently negative: $n$ is the diode ideality factor. The second term gives the photocurrent. Figure 2.5 shows *IV* characteristics both in the dark and under illumination. For this discussion, it is assumed that a load resistor, $R_L$, is connected across the electrodes of the diode. Each value of $R_L$ and operating mode produces a load line on the *IV* characteristics: these are drawn in Fig. 2.5. At zero bias, if a very large value of $R_L$ is used, the output voltage, dropped across $R_L$, is

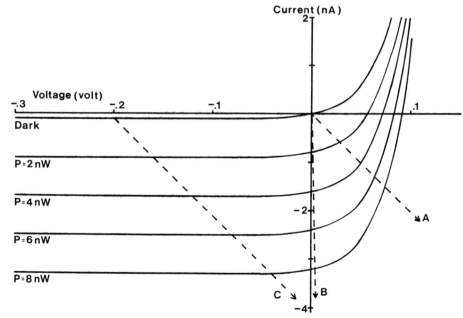

**Fig. 2.5** Current–voltage (*IV*) characteristics of an Si photodiode. P is the radiant power. A, B and C are load lines (see text).

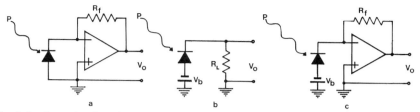

Fig. 2.6  Operational modes of a photodiode: (a) zero bias; (b) reverse bias with load resistor; (c) reverse bias into transimpedance op-amp. P, radiant power; $R_f$, feedback resistor; $R_L$, load resistor; $V_b$, bias voltage; $V_o$, output voltage.

highly non-linear with radiant power, due to internal forward biasing of the junction (load line A). This is known as open circuit mode and must be avoided at all costs. Using a low value of $R_L$ (load line B), the short circuit mode, the output current is linear with radiant power. This is the best way to use a photodiode. The low value load resistance can be obtained by using a low value resistor, when dealing with high radiant powers, or by connecting the diode across the input terminals of an inverting (trans-impedance) operational amplifier. Under reverse bias, similar behaviour is obtained but linearity with a high value of $R_L$ is slightly better (load line C).

In practice there are 3 acceptable modes of operation (Fig. 2.6). Confusion regarding nomenclature has been discussed by Geist (1986). The present author uses "zero-bias operation" for mode (a) and "reverse-bias operation" for modes (b) and (c), differentiating between the latter by "into a transimpedance amplifier" when necessary. The commonly used expressions, photovoltaic for mode (a) and photoconductive for modes (b) and (c), are to be avoided since, in the context of photodiode operation, they are not used correctly.

In measurement of low light levels, zero-bias operation (mode (a)) is simplest since there is no dark current to be subtracted. Under reverse bias (modes (b) and (c)), allowance must be made for the dark current. Since the latter is temperature dependent the method for its subtraction can be complex in DC measurements. The main advantage of reverse-bias operation is the reduction in device capacitance, due to the increased depletion width, and the resultant increase in speed of response. When used with a unity-gain-stable operational amplifier the rise time is given by

$$\tau \approx (R_f C/\omega_a)^{1/2} \tag{14}$$

where $C$ is the parallel sum of the device, amplifier input and feedback capacitances and $\omega_a$ is the gain-bandwidth product of the amplifier. Under reverse-bias operation, there can also be a slight reduction in noise. Mode (b) is normally used when studying fast laser pulses. The rise time is the product of the device capacitance ( < 20 pF) and the load resistance (50 Ω) and is < 1 ns.

Devices are available covering the wavelength range 0.2–12 $\mu$m. These include (figures in parentheses are the spectral ranges in $\mu$m): GaP (0.2–0.55); GaAsP (0.2–0.68); Si (0.2–1.1); Ge (0.7–1.9); InGaAsP (0.6–1.9); InAs (1.0–3.5); InSb (1.0–5.5); HgCdTe (3.0–12.0). The current responsivities of some of these detectors are included in Fig. 2.3. Since $E_g$ has a strong temperature dependence, the responsivity near $\lambda_{co}$ shows a marked change with temperature. Well below $\lambda_{co}$, $\mathcal{R}_i$ shows only a small temperature variation.

Noise in a photodiode can be due to Johnson noise in the zero-bias slope resistance of the device ($R_{d0} = nkT/ei_0$), Johnson noise in the load or feedback resistor, shot noise in the dark current, shot noise in the signal or background photocurrent and amplifier/device-related noise. The dominant noise source depends on the type of photodiode, the modulation frequency and device temperature. With GaAsP and Si photodiodes, Wilson and Lyall (1986a) have shown that amplifier/device noise is significant. Since this noise source is seldom discussed in manufacturers' data sheets, the NEP data therein should be treated with caution. For Si photodiodes NEP values lie in the range $10^{-15}$—$10^{-12}$ W.Hz$^{-1/2}$. With an infra-red photodiode such as InSb cooled to its normal operating temperature of 77 K, Zhang and Williamson (1982) have shown that both Johnson noise in $R_{d0}$ and shot noise in the background photocurrent are dominant.

The linearity of a photodiode is limited at high currents by the effects of resistance in series with the diode. For many silicon devices, non-linearity becomes significant at photocurrents of 1 mA. A dynamic range of $10^{10}$ is possible with GaAsP and Si photodiodes (Wilson and Lyall, 1986a).

The photodiodes discussed till now have had unity gain. However, if a p–n junction is operated under very high reverse-bias, the photogenerated carriers, in being accelerated across the depletion region, can acquire sufficient energy to promote additional electrons to the conduction band by impact ionization. This is the avalanche photodiode (Murray et al., 1980) and it can be considered as a solid state analogue of the photomultiplier. Internal gains of about 100 are normally obtained. This is useful where a fast response to low light levels is required. Gain can also be obtained using a phototransistor in which the base-collector junction is left exposed. Devices are available both with and without the base connection accessible. The former are preferable since appropriate base biasing can be used to stabilize the operating point and reduce the otherwise significant dependence of the gain on the input radiant power. Since carriers have to diffuse across the base-collector junction, the response time is modest (5 $\mu$s or more). For accurate radiometry, conventional photodiodes will always be superior.


### 2.4.4. Photoconductors

A photoconductive detector is a slab of semiconductor material furnished with two electrodes which provide ohmic electrical contacts (Sze, 1969). When illuminated by light of an appropriate wavelength free carriers are generated in the semiconductor thereby changing the conductivity and resistance of the slab. Either intrinsic or extrinsic semiconductors can be used. In intrinsic detectors, the photon energy must exceed the band gap energy of the semiconductor, $(h\nu > E_g)$. Examples include (the figures in parentheses give the spectral range in $\mu$m): CdS (0.4–0.58); CdSe (0.5–0.78); PbS (1.0–3.0); PbSe (1.0–6.0); InAs (1.0–3.5); InSb (1.0–5.5); HgCdTe (3.0–12.0). In extrinsic detectors, the photon excites a carrier to or from a forbidden gap energy level associated with dopants. In practice these transitions require very small energies and thus extrinsic detectors have responses in the infra-red and far infra-red. Examples include: Ge:Hg (2.5–15); Si:As (5–25); Ge:Be (10–50); Ge:Ga (20–120).

A photoconductive detector is commonly used in the circuit of Fig. 2.7.

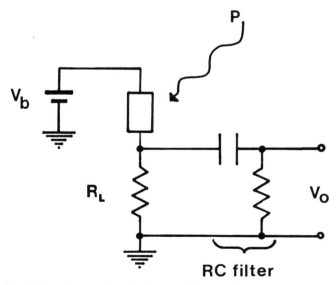

**Fig. 2.7** Circuit for a photoconductive detector including a high pass (RC) filter. P, radiant power, $R_l$, load resistor; $V_b$, bias voltage; $V_o$, output voltage.

An applied bias voltage, $V_B$, causes a bias current to flow even in the dark. The voltage dropped across the load resistor, $R_L$, is given by

$$V = V_B R_L/(R_L + R_d) \tag{15}$$

where $R_d$ is the resistance of the detector. When illuminated, the free carriers reduce the value of $R_d$ and thus $V$ is altered. It is normal practice to modulate the light source and to capacitively couple (high pass filter) $V$ to the amplifier or measuring circuit. This eliminates DC effects including that due to the biasing and leaves only the AC signal voltage, $\Delta V$, due to the radiation: $\Delta V$ is given by

$$\Delta V = \mathscr{R}_i P R_L R_d/(R_L + R_d) \tag{16}$$

where $\mathscr{R}_i$ is the current responsivity and $P$ is the radiant power on the detector. Note that $\Delta V/P$ is simply the voltage responsivity. Other biasing arrangements can be used depending upon the value of $R_d$ and type of application (Jarratt, 1971): amplifier circuits are also discussed in this reference.

The current responsivity of a photoconductor is given by

$$\mathscr{R}_i = \left(\frac{e\eta\lambda}{hc}\right) G \tag{17}$$

Where $G$ is the photoconductive gain. After photogeneration of an electron-hole pair, the minority carrier may be trapped at impurity or defect sites. The majority carrier drifts out of the slab at one contact and to maintain charge neutrality another majority carrier is injected at the other contact. This drift/injection process continues until a majority carrier recombines with the minority carrier. This increases the majority carrier lifetime, $\tau_L$, and results in gain because more than one majority carrier flows in the external circuit for every photogenerated carrier. While this process does lead to very high values for the current responsivity it has a major disadvantage since the response time of the detector, which is approximately equal to $\tau_L$, is also increased. Response times vary from 100 ms for CdS (a high gain material) to sub-microsecond for HgCdTe detectors.

The major noise sources in a photoconductive detector include $1/f$ noise, generation–recombination noise and Johnson noise. The $1/f$ noise is only significant at low frequencies and can be eliminated by operating at a sufficiently high modulation frequency. At intermediate frequencies up to the reciprocal of the carrier lifetime, the flat noise spectrum of generation–recombination noise is often dominant. This noise may originate in the dark current, the background photocurrent or the signal photocurrent. IR photoconductors are sensitive to radiation emitted from surfaces at near-ambient temperatures: a black body at 300 K has a peak spectral emission at 10 $\mu$m. The background current is often larger than the dark current and thus noise is dominated by the former. This is referred to as the background-limited infra-red photodetector, or BLIP (Dereniak and Crowe, 1984). At high frequencies, above the carrier lifetime roll-off, Johnson noise in the detector and load resistor dominates. A typical NEP value for a PbS detector is $10^{-10}$ watts.

From Equation (16) it is clear that $\Delta V$ depends on $R_d$ in a non-linear fashion since $R_d$ occurs in the denominator. It is still possible to get linear operation over 3–4 orders of magnitude with PbS detectors.

## 2.5. Photodetector Formats

Photodetectors are available in a wide variety of formats and sizes. Photomultipliers can have photocathode diameters from < 1 cm to over 30 cm. Solid state photodetectors are available with active elements ranging in size from 20 $\mu$m to 20 mm. Infra-red (IR) photoconductors, such as PbS, are available with Peltier thermoelectric coolers integrated into a TO-type package to allow operation at the optimum low temperature for highest $D^*$. Other IR-sensitive detectors, such as InSb photoconductors and photodiodes, are normally supplied already mounted in a liquid nitrogen dewar. Some IR detectors are provided with immersion lenses to increase their effective area or with lens assemblies which define their field of view. Many detectors are supplied with pre-amplifiers integrated into a TO-type package: these are useful for routine measurements, but for ultra-low noise applications, it can be better to use discrete components.

Multi-element detectors are also available. A common format for Si photodiodes and IR photon detectors is 4 discrete elements forming the quadrants of a circle, connected such that there is an output signal only when they are illuminated unequally. They are used for position sensing and are capable of sub-micron resolution. The ultimate extension is the detector array, which may use linear or rectangular arrays of Si photodiodes, pyroelectric detectors or infra-red photoconductors, such as HgCdTe. They are used for a variety of tasks including imaging, as in astronomy and thermal imaging cameras, spatial profiling of laser beams and in optical multi-channel analysers.

## 2.6. Example Applications

From the previous sections it is clear that GaP, GaAsP and Si photodiodes and photomultipliers are of great use in the spectral region below 1.1 $\mu$m. For a general laboratory radiometer required to measure nW to mW power levels with a bandwidth of a few kHz, a Si photodiode operated at zero-bias into a JFET op-amp is suitable and easy to make and use. Often it will be necessary to use optical filtering to restrict the detector's spectral response to, say, a narrow band in the UV. Here GaP and GaAsP photodiodes and certain photomultipliers can offer major gains in performance over Si devices since their much lower cut-off wavelengths reduce the requirements on the filters to block out-of-band radiation. For measuring low light levels ($<$ nW) at high bandwidths (MHz), photomultipliers are most useful. Vacuum phototubes and reverse-biased Si pin photodiodes are commonly used for measuring the temporal profiles of laser pulses.

In the infra-red (IR) region (1.1—12 $\mu$m) the choice of photon detectors is wide, each type being dominant over a relatively narrow waveband. For example, a PbS photoconductor is probably the best choice for a detector at around 2 $\mu$m. However, a thermal detector, such as the pyroelectric, may be suitable for many IR measurements. In view of its simplicity of use (no cooling required, robust, small size), pyroelectrics are always worth invest-igating before considering use of IR photon detectors. Pyroelectrics are certainly very good for measurement of the energy and power output of IR lasers.

In some instances thermal detectors will be the only choice. To measure solar radiation from 0.2 to 3.0 $\mu$m using a single detector is a case in point. Many commercial pyrheliometers (the name for this type of instrument) use thermopiles. A pyroelectric detector could also be used if the radiation can be chopped. For both types of detector there is ample irradiance (95 mW.cm$^{-2}$ for the Air Mass 1 emission) for good signal to noise ratios. The main design consideration is the uniformity and durability of the black coating used to obtain a response over the required spectral band.

Finally, consider the difficult task of monitoring long term (hours to days) exposure to UV radiation. One solution is to use a UV-stable, UV-sensitive photodiode (GaAsP) with an integrator which resets when a specified charge has been accumulated, each reset being registered using a digital counter. Alternatively, use could be made of the bleaching of dyes by UV: the dye acts as a somewhat unconventional photon detector. Before using a dye detector it would be necessary to determine its spectral response (action spectrum), effective responsivity (change in absorptance, $\Delta\alpha$, per unit energy of UV radiation) and its linearity (variation of $\Delta\alpha$ with UV dose). These are, of course, some of the parameters that it would be necessary to determine for any detector.

# References

Blevin, W. R. & Geist, J. (1974). Influence of black coatings on pyroelectric detectors. *Appl. Opt.* **13**, 1171–1178.

Chiari, J. A. & Morten, F. D. (1982). Detectors for thermal imaging. *Electronic Comp. Appl.* **4**, 242–252.

Dereniak, E. L. & Crowe, D. G. (1984). "Optical Radiation Detectors", pp. 36–59. Wiley, New York.

Geist, J. (1986). Photodiode operating mode nomenclature. *Appl. Opt.* **25**, 2033–2034.

Geist, J. & Zalewski, E. F. (1979). The quantum yield of silicon in the visible. *Appl. Phys. Lett.* **35**, 503–506.

Glass, A. M. & Abrams, R. L. (1970). High frequency performance of pyroelectric detectors. *In* "Submillimeter Waves" (J. Fox, Ed.), pp. 281–294. Polytechnic Press, New York.

Golay, M. J. E. (1952) Bridges across the infrared-radio gap. *Proc. IRE* **40**, 1161–1165.

Hamamatsu Co. Ltd (1979). "Photomultiplier Tubes".

Hamstra, R. H. & Wendland, P. (1972). Noise and frequency response of Silicon photodiode operational amplifier combination. *Appl. Opt.* **11**, 1539–1547.

Horowitz, P. & Hill, W. (1980). "The Art of Electronics", pp. 286–313. Cambridge University Press, Cambridge.

Jarratt, T. J. (1971). Biasing and amplifying techniques for photoconductive detectors'. *In* "Applications of Infrared Detectors" (F. A. Sowan, Ed.), pp. 26–44. Mullard, London.

Meade, M. L. (1981). Instrumentation aspects of photon counting applied to photometry'. *J. Phys. E: Sci. Instrum.* **14**, 909–918.

Murray, L. A., Wang, K. & Hesse, K. (1980). A review of avalanche photodiodes, trends and markets. *Opt. Spectra.* (Apr.), 54–59.

Rose, M. A. (1982). Pyroelectric infrared detectors. *Electronic Comp. Appl.* **4**, 142–149.

Satheeshkumar, M. K. & Vallabhan, C. P. G. (1985). Use of a photoacoustic cell as a sensitive laser power meter. *J. Phys. E: Sci. Instrum.* **18**, 434–436.

Seib, D. H. & Aukerman, L. W. (1973). Photodetectors for the 0.1 to 1.0 $\mu$m spectral region. *In* "Advances in Electronics and Electron Physics", (L. Marton, Ed.), Vol. 34, pp. 95–221. Academic Press, New York.

Smith, R. A., Jones, F. E. & Chasmar, R. P. (1957) "The Detection and Measurement of Infra-Red Radiation", pp. 204–207. Oxford University Press, Oxford.

Sze, S. M. (1969). "Physics of Semiconductor Devices", pp. 625–686. Wiley, New York.

Thorn-EMI Electron Tubes Ltd (1986). "Photomultipliers".

Wilson, A. D. & Lyall, H. (1986a). Design of an ultraviolet radiometer. 1: Detector electrical characteristics. *Appl. Opt.* **25**, 4530–4539.

Wilson, A. D. & Lyall, H. (1986b). Design of an ultraviolet radiometer. 2: Detector optical characteristics. *Appl. Opt.* **25**, 4540–4546.

Zhang, Y-X. & Williamson, F. O. (1982). Evaluation of an InSb infrared detector at liquid $N_2$ and liquid He temperatures. *Appl. Opt.* **21**, 2036–2040.

# 3
# Calibration of Light Sources and Detectors

## T. M. GOODMAN

*National Physical Laboratory*

The sources and detectors used as standards for the measurement of optical radiation (i.e. visible radiation and the adjacent spectral regions of the ultraviolet and infra-red) are the end products of a calibration chain which begins with the primary standards held at the national laboratories. Although there will be uncertainties associated with the values assigned to these secondary standards (and these should be stated on the calibration certificate) these are often small compared with the errors that can be introduced if the standard itself, or the calibration data supplied with it, are used incorrectly.

The purpose of the present chapter is to examine the use of secondary standard sources and detectors in a number of the more common measurement applications, and to highlight some of the possible sources of error that can arise. The subject is treated, throughout, from the point of view of an experimental scientist using a working standard lamp or detector, which has been calibrated by a standards laboratory, to make measurements on other sources. The establishment of the primary standards themselves and the methods used by standardizing laboratories to calibrate the working standards are not, therefore, discussed in any detail.

## 3.1 Calibration of Light Sources

The term "light source" covers a wide variety of emitters of optical radiation, from the sun, to mercury discharge lamps, to light-emitting diodes (LEDs) and lasers, and covering the wavelength range from the ultraviolet to the near infra-red. As might be expected, a considerable diversity also exists in the concepts used to describe the output, in terms of both the geometry of collection of the radiation and the way in which this radiation is evaluated (as described in Chapter 1). It is particularly important to note the distinction between radiometric quantities, where the radiation is evaluated in purely physical terms (i.e. in terms of the power) and photometric quantities, where it is weighted by means of a standard photometric observer.

There are two general methods by which a test source can be measured. One of these involves a direct measurement using a calibrated detector, and this method is discussed further in Section 3.2 below. The other method, which is treated here, involves a comparison with a calibrated reference source and, in this case, the detector acts simply as a transfer device.

Measurements can be made either "broadband", using a detector with a spectral responsivity tailored to correspond as closely as possible to a given response function, or spectrally, i.e. in terms of the power in narrow wavebands across the spectral region of interest. The most common example of the former case is a photometer designed to simulate the $V(\lambda)$ function (see Chapter 1) on which photometry is based and which corresponds to the internationally agreed curve for the spectral sensitivity of the average human eye. Detectors are also available, however, which attempt to match other biological weighting functions, such as cutaneous erythema. With broadband measurements, each detector/filter combination can only represent a single weighting function, but a spectral measurement on a source enables the effect of a variety of different weighting functions to be investigated. Spectral measurements may also be more appropriate if the required responsivity function cannot be accurately matched by the use of a filtered detector.

Tungsten filament lamps are often the most suitable choice as calibration standards. They are very stable and reproducible, have low calibration uncertainties, are relatively simple to operate and are available in many different power ratings, sizes and shapes. They can be calibrated for use as spectral or broadband standards, for viewing in a specified direction or integrated over the full solid angle of $4\pi$ steradians. However, there are many situations where such a standard bears little relation to the test sources to be measured (for example, the spectral power distribution may be different, as with a fluorescent lamp) and here the properties of the transfer detector can have a major influence on the results obtained. In many such cases, it may be advisable to choose a standard with characteristics similar to those of the test source.

Thus the choice of a suitable calibration standard will be determined not only by the quantity to be measured, but also by the properties of the available standards, the test sources to be measured and the level of uncertainty which is required.

### 3.1.1. Luminous intensity and illuminance

These quantities relate to the directional properties of a source: the luminous (or radiant) intensity characterizes the flux from the source in a specified direction while the illuminance (or irradiance) gives the flux incident on a surface at a given distance from the source in the specified direction. For a point source the two concepts are related through the inverse square and cosine laws:

$$E = I \cos \theta / d^2$$

where $E$ is the illuminance or irradiance, $I$ is the luminous, or radiant, intensity, $d$ is the distance from the source to the irradiated surface and $\theta$ is the angle between the normal to the surface and the direction of the incident radiation.

The terms radiant intensity and irradiance are used if measurements are made on a radiometric or power basis. Spectral power measurements relate to the spectral concentration of intensity or irradiance and the units are watts per steradian per nanometre or watts per square metre per nanometre, respectively. Such measurements will be discussed further in Section 3.5.1. However, if the measurements are made in terms of the responsivity of the average human eye, as defined by the internationally agreed $V(\lambda)$ function, then the terms used are the luminous intensity and illuminance, given in candelas and lux, respectively.

The most commonly used standard source for luminous intensity and illuminance calibrations is the tungsten filament lamp. The transfer detector is a $V(\lambda)$-corrected photometer, often with a diffusing disc or dome attached to the front. The latter gives an angular responsivity approximating to the ideal cosine law responsivity function $s(\theta) = s(0)\cos \theta$ where $s(\theta)$ is the responsivity at an angle of $\theta$ to the normal.

Calibrations are generally performed on a photometric bench, which enables the source and the measuring head (photometer) to be arranged on the same axis and at a known distance apart. The bench should be situated in a dark room or at least be screened from the surroundings by thick black curtains. The walls, ceiling and floor should all be blackened, to reduce unwanted reflections, and screens (also matt black) should be placed at intervals along the bench, to prevent stray light from reaching the photometer (Figure 3.1). The bench is usually fitted with a scale, to allow accurate measurement of the distance between the source and detector, and a telescope is mounted perpendicular to the optical axis to allow the lamp filament to be viewed and positioned exactly, in relation to the bench scale.

There are two methods by which a test source can be calibrated against a reference source of known luminous intensity. The first involves direct substitution of the test lamp for the reference lamp, ensuring that the separation between the photometer and the lamp is kept constant. The signals produced by the photometer when exposed to the reference and to the test source are $x_r$ and $x_t$, respectively. The ratio of these signals is the

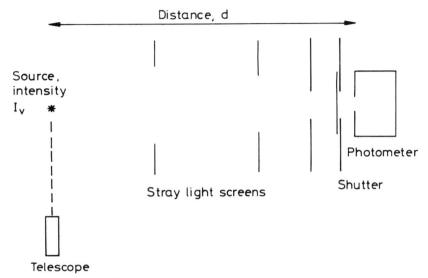

**Fig. 3.1** Photometric bench (not to scale).

same as the ratio of the luminous intensities ($I_r$ and $I_t$), giving $I_t = I_r x_t / x_r$. The second method involves adjusting the distances $d_r$ and $d_t$ between the lamp and the photometer to generate an identical output in each case. The luminous intensities are then in the ratio of the distances squared, so $I_t = I_r (d_t / d_r)^2$. The first method is generally easier to implement, but the second can be used with photometers with a non-linear output (i.e. an output which is not proportional to the input) provided that the sources obey the inverse square law. In general, however, non-linear detectors can and should be avoided.

An illuminance standard is usually a lamp which has been calibrated for luminous intensity placed at a known distance from the photometer. If the photometer is fitted with a diffuser, then the distance is measured from the front surface of this diffuser. If no diffuser is fitted, however, the distance must be measured from the position of the limiting aperture of the photometer. In the latter case, care must be taken to correctly identify the limiting aperture position and if it lies behind any glass components (e.g. behind the $V(\lambda)$ filter) correct allowance must be made for the optical thickness of these components. Having correctly determined the distance, $d$, and knowing the luminous intensity, $I_v$, the illuminance can be calculated:

$$E_v = \frac{I_v}{d^2}.$$

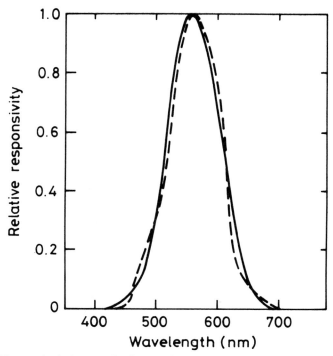

**Fig. 3.2**  The standard photometric observer function ($V(\lambda)$——) and the spectral responsivity curve for a high quality photometer ($s(\lambda)$- - -).

The illuminance at the detector can be varied by a known amount by changing the distance, provided the lamp obeys the inverse square law.

A number of points should be noted in connection with both types of calibration.

(i) Standard lamps are always calibrated under specified alignment conditions and for viewing in a stated direction. These conditions should be accurately reproduced each time the standard is used.

(ii) No photometer perfectly matches the $V(\lambda)$ function, but will have areas of mismatch between $V(\lambda)$ and the photometer responsivity (see Fig. 3.2). Hence, if the source to be measured has a spectral characteristic which differs significantly from that of the reference source, an error can arise. A correction, called the colour correction factor, can be calculated to allow for the photometer mismatch, but this requires knowledge of the spectral power distributions of both the test and reference sources ($S_t(\lambda)$ and $S_r(\lambda)$) and also the spectral responsivity of the photometer ($s(\lambda)$):

$$F = \frac{\int S_t(\lambda) V(\lambda) d\lambda}{\int S_t(\lambda) s(\lambda) d\lambda} \times \frac{\int S_r(\lambda) s(\lambda) d\lambda}{\int S_r(\lambda) V(\lambda) d\lambda}.$$

In many circumstances where this information is not available a systematic error will result unless the reference source is chosen to have a spectral power distribution which is similar to that of the source which is to be measured.

An additional complication can arise when measurements are made on a source *in situ*. Any significant coloration of the surroundings (e.g. the walls or ceiling) will result in a corresponding coloration of the ambient illumination. So in this case, any correction which is applied to allow for spectral mismatch between the photometer and $V(\lambda)$ will have to take account not only of the source characteristics but also the surroundings.

(iii) For a point source, the photometer reading or illuminance is proportional to $I/d^2$ but in practice most sources obey this inverse square law only imperfectly, particularly at short distances. Standard lamps are calibrated at a stated distance and if errors are to be avoided, compliance with the inverse square law should be checked before using the standard at any other distance, particularly at very short lamp to photometer separations.

(iv) For sources which do not radiate uniformly in all directions, the measured luminous intensity will vary with the solid angle over which it is collected. Generally, this effect is small, but for highly directional sources, such as lensed LEDs, it can lead to significant errors.

(v) An ideal illuminance meter would have a perfect cosine response $s(\theta) = s(0)\cos\theta$ where $s(\theta)$ is the responsivity at angle $\theta$ to the normal. Practical meters never perfectly achieve such an ideal and this can lead to errors when an instrument calibrated with a standard at $0°$ is used to measure the illuminance in, say, a room where light is incident from many directions.

(vi) Stray light errors can be appreciable, particularly when comparing sources of very different size or shape. The use of screens between the source and photometer as described previously is therefore essential. A simple check for stray light is to place a baffle directly between the source and the photometer, so as to obstruct all direct illumination. Any signal then recorded is due to stray or scattered radiation and can either be deducted from the readings or, preferably, eliminated by more careful screening.

(vii) Although standard lamps are specifically designed to be stable and reproducible, they will inevitably age with use. If extensive running of the standard is envisaged, it is generally recommended that a working group of lamps be calibrated against the reference standard, and used for routine work. These working standards can be checked regularly against the reference, and their values adjusted as they age.

## 3.1.2. Luminous flux

The luminous intensity of a source gives its output in a specified direction, but it is often necessary to know the total luminous output emitted into the full solid angle of $4\pi$ steradians, i.e. the total luminous flux. The luminous flux scale is maintained at the national laboratories in groups of tungsten filament lamps and is derived directly from the luminous intensity scale.

The luminous flux of a source can be measured by two methods. The first uses a goniophotometer; this is essentially a photometer fixed at one end of an arm which pivots about the light centre of the lamp, enabling the intensity at each point around the surface of the lamp to be measured. The luminous flux is then given by the mean luminous intensity multiplied by $4\pi$. The photometer is calibrated using luminous intensity standards as on a photometric bench and this method therefore links the lumen directly to the candela. This is the technique used by national standards laboratories to derive their scale of luminous flux from the luminous intensity scale.

The second method uses an integrating sphere coated with a white diffusing paint (Fig. 3.3) for which it can be shown that the illuminance at any point on the surface of the sphere is directly proportional to the luminous flux of a source placed within it (Walsh, 1965). This technique requires the use of a standard lamp of known luminous flux to calibrate the sphere; the ratio of the fluxes is then given by the ratio of the photometer

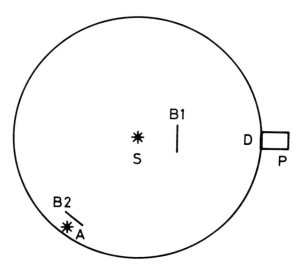

**Fig. 3.3** Integrating sphere for the measurement of luminous flux (not to scale). A, auxiliary lamp; B1, B2, screens; D, diffusing window; P, photometer; S, source.

readings. It should be noted that while the theory only truly applies for a spherical integrator, other shapes, such as cubes, can give a sufficiently good approximation for most purposes.

In general, the integrating sphere method is the quickest and simplest to use, but the goniophotometer is more appropriate in situations where information on the spatial distribution of the flux is also required. Both methods can be subject to error, some of the most significant being the following.

(i) As in the measurement of luminous intensity and illuminance, mismatch between the photometer responsivity and the $V(\lambda)$ function can be a major source of error. As a general rule, the reference lamp should be of a similar type to the test lamp, to minimize this effect. If this is not possible then a correction can be applied, provided that information on the spectral characteristics of the photometer and the sources is available. When calculating the correction required for measurements made in an integrating sphere, allowance must also be made for any coloration of the sphere paint. This coloration will be enhanced by the repeated reflections within the sphere and for this reason the paint should be as non-selective as possible. Often, the reflectance of the paint is reduced by the addition of carbon black to give a reflection factor of about 80%, thus reducing the number of reflections within the sphere and hence improving the neutrality.

(ii) For a goniophotometric measurement, the optical pathlength should be at least five times the largest dimension of the source being measured if breakdown of the inverse square law is not to lead to significant errors.

When using an integrating sphere, the substitution principle only strictly applies if the size, shape and directional output of the two sources being compared are the same. If the sources are not identical, then the systematic errors due to imperfections in the paint and the presence of the screen (which is required to prevent direct illumination from reaching the photo-meter) will be different in the two cases, and will no longer cancel. However, provided that the sphere is large compared with the source (diameter at least three times the largest dimension of the source) and the screen is kept small (just large enough to obstruct direct illumination) these errors will be negligible.

(iii) Any body placed within an integrating sphere (including the lamp itself) will absorb some of the radiation emitted by the source. When comparing two physically identical lamps, the amount of radiation absorbed by each will be the same. If the two lamps are not identical, for example if one has a blackened bulb compared with the other or is of a different size, shape or type, then a self-absorption error can be introduced. A correction factor for this error can be obtained by the use of an auxiliary lamp (marked A on Fig. 3.3) placed close to the sphere wall and screened so as to prevent any light from it falling directly onto the photometer or the source being measured. With the auxiliary lamp lit, readings are taken first with the standard lamp in position but unlit (reading $R_s$) and then with the test lamp in position but again unlit (reading $R_t$). The ratio $R_s/R_t$ is then used to correct readings taken in the usual way.

A special type of self-absorption error arises with certain gas discharge lamps (such as low-pressure sodium lamps) where radiation at the emission lines is strongly absorbed when the lamp is running but not when it is cold. In this case, measurements made using the auxiliary lamp method cannot be used to correct the readings.

(iv) Errors due to non-uniformity of the paint reflection factor within an integrator are difficult to quantify and can be large for sources with a non-symmetrical distribution. It is necessary, therefore, to keep the sphere clean and repaint it at fairly frequent intervals, since dirt and dust collect chiefly on the lower half.

(v) Excessive use of standard lamps can lead to large changes in output due to ageing and should therefore be avoided. As in the case of luminous intensity measurements, working standards can be set up and used for routine work.

### 3.1.3. Luminance

The luminance $L_v$ of a surface is defined as the luminous intensity divided by the projected area of the surface in the direction of view and is measured in candelas per square metre. It characterizes the luminous flux in a specific direction and at a specific point on the surface of a source, and is related to the luminous intensity by the equation:

$$L_v = \frac{\mathrm{d}I_v}{\mathrm{d}A \cos \theta}$$

where $\mathrm{d}A$ is the area of the element of surface and $\theta$ is the angle between the normal to this element and the specified direction.

A luminance meter generally consists of a photometer with an imaging system to focus the area being measured onto the detector. The optics are usually designed to allow the area being measured to be viewed and identified through an eyepiece. There is also some form of diaphragm to isolate the area being measured and this is often adjustable to enable fields of different sizes (typically in the range $1'$ to $3°$) to be examined.

The measured luminance of most sources varies with the collection angle used and the position on the surface of the area viewed. A so-called Lambertian source, however, has a luminance independent of the angle from which it is viewed and therefore independent also of the collection angle of the meter. Such sources are preferred as luminance standards. A piece of white opal glass illuminated at normal incidence using a luminous intensity standard lamp and viewed at an angle of 45° to the normal provides a close approximation to a Lambertian source at that particular angle of view. If the luminance factor (the reflectance) of the opal is known, then this gives an absolute luminance standard, which is not only approximately Lambertian, but is also very uniform across its surface. Luminance standards of this type are related directly to the luminous intensity scale, but are often somewhat inconvenient to use. Consequently a variety of luminance gauges operating on a different principle have been developed for routine use. These generally consist of a lamp with an integrating sphere or diffuser to generate the desired uniform Lambertian field (see Fig. 3.4). Many gauges of this type can provide a variable level of luminance, usually by means of an adjustable aperture, and they are very useful for checking the performance of a meter across its working range.

As in the case of other photometric measurements, spectral mismatch between the luminance meter responsivity and the $V(\lambda)$ function means that colour differences between the reference and test sources can lead to significant error. Colour correction factors may have to be applied, particularly when measuring highly coloured sources.

In most cases, luminance gauges are designed to provide a radiant field at a colour temperature of 2856 K (source A in the CIE (Commission Internationale de l'Éclairage) nomenclature) but the presence of the sphere or diffuser can result in significant departures from a tungsten or black body spectral distribution, particularly in the blue region. If a colour correction factor is calculated, it is not sufficient to assume a black body distribution;

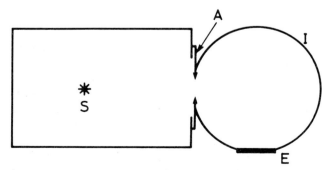

**Fig. 3.4**  A variable luminance gauge (not to scale). A, entrance port with adjustable aperture; E, exit port giving uniform Lambertian field; I, integrating sphere; S, lamp in enclosure.

the computation should always be based on actual measurements of the spectral characteristics of the gauge.

Some luminance meters have an additional "close-up" lens which can be affixed to enable very small areas to be measured. However, these are frequently anti-reflection coated and have the effect of colouring the light incident on the photometer. Although this point is often not mentioned in the manufacturers' instructions, it may be necessary to apply an additional colour correction factor to allow for this coloration. Even a spectrally neutral lens will change the overall responsivity of the instrument, so unless it has been calibrated with the lens in position, an appropriate correction should be applied.

Sources not specifically designed as luminance standards generally show a variation in luminance across their surface, so if these are to be measured, the exact area being examined should be readily identifiable. Masking of the surface to isolate a particular area may be appropriate in such cases but in any event, the shape and size of the calibrated area should be specified. In addition, since the majority of sources are not perfectly Lambertian, i.e. they show a variation in luminance with the direction of view, it is important to record the conditions of alignment between the source and the meter, and the field angle over which the measurements are made. With a highly directional source, measurements at different viewing or collection angles can give results differing by many tens of per cent.

## 3.1.4. Low-level photometry

Particular problems are associated with the calibration of low-level sources and extra care is needed if errors are to be avoided. Before discussing these, however, a note on units is required. Photometric measurements are made in terms of the $V(\lambda)$ curve, which represents the spectral responsivity of the standard human eye under normal lighting levels (photopic vision). But the spectral sensitivity of the eye actually changes as the illumination level is reduced, and at very low levels is represented by the scotopic curve $V'(\lambda)$ (see Fig. 3.5). This latter curve may be more appropriate therefore for the measurement of low-level sources, and in any such calibration, the nature of the weighting curve used should be specified.

Those problems already discussed in connection with measurements on more general sources (e.g. mismatch between the photometer responsivity characteristics and $V(\lambda)$) apply in equal force to low level sources. In addition, the following points should be checked.

(i) If comparing a low-level source with a much brighter standard, non-linearity of the detector can introduce errors. If possible, the standard

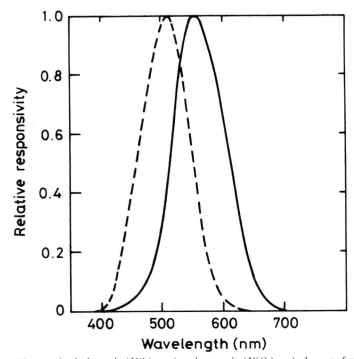

**Fig. 3.5**   The standard photopic ($V(\lambda)$——) and scotopic ($V'(\lambda)$- - -) observer functions.

should be of similar output to the test or the method used should be such that detector non-linearities do not influence the result. An alternative approach is to check the linearity of the detector using one of the standard techniques (Saunders, 1972).

(ii) A related source of error can arise in changing the range of the photometer. Most devices allow the output to be increased in steps of 10:1, but often this ratio is not precise. The relationship between the ranges can be checked simply by comparing the readings obtained on different ranges with a source of stable output.

(iii) It is often tempting to bring the source very close to the detector in order to increase the signal, but this can lead to errors due to departure from inverse square law behaviour.

(iv) At low levels it becomes particularly important to make correct allowance for stray light and for the detector dark signal. Careful screening and correct positioning of the shutter are essential.

(v) Signal-to-noise considerations can become important, and may require the use of a different type of detector. For example, photomultiplier detectors have a higher responsivity than solid state detectors.

### 3.1.5. Spectroradiometry

Spectroradiometric measurements give the distribution of the energy from a source as a function of wavelength and can cover not only the visible but also the ultraviolet and infra-red regions. Spectral power distribution is a particularly important parameter when assessing the biological effects of a source (e.g. erythema, photosynthesis). If an appropriate action spectrum is known, then it can be combined with a measurement of the absolute spectral power distribution to evaluate the biological effect in much the same way as the $V(\lambda)$ function enables the visual sensation to be assessed.

The primary spectroradiometric standard may be an absolute source, such as a black body or synchrotron radiator, whose spectral characteristics can be calculated from established physical laws. Alternatively it may consist of an absolute detector of known spectral responsivity, such as an electrically calibrated radiometer. Such base standards are usually maintained at the national laboratories and are used to establish groups of suitable secondary standards which then serve to maintain and disseminate the scale.

Various sources are available for calibration as secondary standards of spectral power distribution and the most appropriate will depend on the spectral region to be studied. For example, a tungsten lamp will not be suitable for the UVB and C regions, since the proportion of its energy emitted at these wavelengths is very small. A discharge source, such as a deuterium lamp, would be a better choice. On the other hand, tungsten lamps are ideal for use in the visible and near infra-red, due to their stability, reproducibility and ease of use.

A typical system used for making spectroradiometric measurements consists of an adjustable device (often a monochromator), which isolates a single narrow wavelength band from the source, and a suitable detector (Fig. 3.6). This allows the outputs of a test and a reference source to be compared, usually by scanning the monochromator across the wavelength region of interest with the reference source in position, thus calibrating the system, and then repeating the scan with the test source.

The monochromator is the central element of the system but the characteristics of all the different components can have a significant effect on the accuracy of the results. The points discussed below are particularly

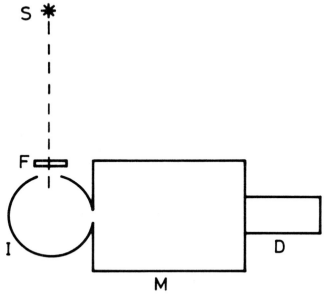

**Fig. 3.6**  The measurement of spectral power distribution (not to scale). D, detector; F, second-order filter; I, integrating sphere; M, monochromator; S, source.

important; more details can be obtained from Moore (1980) or from the CIE (1984).

The transmission of a monochromator is dependent on the way in which it is illuminated by the source. Some form of input optics is therefore required to ensure that the radiation from both the test and reference sources follows exactly the same optical path through the system. Commonly, a sphere or a plane opal diffuser (irradiated normally and viewed at 45°) is placed between the source and the monochromator. Besides making sure that the entrance slit is illuminated in the same way by different sources, such an arrangement has the added advantage that the radiation is depolarized, thus eliminating another possible source of error (monochromators can show a polarization dependence of up to 40%).

The monochromator bandwidth and the method used to sample the spectrum can have a major influence on the results obtained. Some systems use a stepwise procedure, whereby the monochromator will scan to a wavelength, stop, make a measurement, then scan to the next wavelength, and so on. In this situation it is important that the bandwidth and the step interval are matched, so that no information is lost in the "gaps" between measurements, or included in two consecutive measurements. Other systems use a continuous scanning method, sampling the output throughout a steady scan across the spectrum. A third technique involves the use of a diode array detector placed at the exit slit of the monochromator such that a large segment of the spectrum is sampled simultaneously. For the latter systems, the resolution and the length of the spectrum sampled depends on the bandwidth and dispersion of the monochromator, and the number of elements in the array. They are particularly useful for measurements on sources which vary with time but generally have poorer resolution than scanned monochromator systems due to cross-talk between adjacent elements or loss of information in the dead space between them.

Stray light can be a major problem in spectral measurements, particularly in the blue region of the spectrum, because the radiation in any individual wavelength band is such a small proportion of the total radiation present. It is important, therefore, to check the stray characteristics of the mono-chromator. One of the most straightforward methods for doing this is to use a series of sharp cut-on filters which only pass radiation above a certain wavelength. The monochromator is set to a given wavelength and a filter which cuts on above this wavelength is placed between the source and the entrance slit. Any signal then recorded by the detector must be due to stray light. By using a series of such filters and adjusting the monochromator wavelength, the stray light characteristics can be checked across the whole of the spectral region which is to be sampled. As a general rule, double monochromators have better stray light properties than single mono-chromators and are therefore to be preferred for spectral power measure-ments.

In monochromator systems where the dispersing element is a grating, care must be taken to remove the higher order spectra. A cut-on filter is usually placed between the source and the entrance slit to absorb the unwanted radiation at shorter wavelengths.

The ratio between the signals from, say, a tungsten lamp and from a fluorescent lamp at the peak of one of the line emissions can be very large, and this can lead to errors due to non-linearity of the detector or its associated electronics. The linearity range of the instrument should there-fore be determined, and care taken not to exceed the maximum permissible output signals. This may require the use of attenuating filters, and if this is the case, these should ideally be used with both the reference and test sources. Alternatively, the spectral transmission of the filters can be measured and allowed for in calculating the spectral power of the unknown source.

The data obtained from a spectral measurement can be used to calculate such quantities as the chromaticity, colour rendering or erythemal effect. The principle involved in all these calculations is the same. The spectral power values are grouped into appropriate bands, usually 5 or 10 nm wide,

**Table 3.1.** The calculation of the $X$ tristimulus value for a source involves multiplying the spectral power distribution $S(\lambda)$ at each wavelength by the value of the $\bar{x}(\lambda)$ function at that wavelength and summing the results. The same procedure can also be used with other appropriate weighting functions to calculate such quantities as colour rendering indices or erythemal effect

| $\lambda$ | $S(\lambda)$ | $\bar{x}(\lambda)$ | $S(\lambda)\bar{x}(\lambda)$ |
|---|---|---|---|
| 380 | 9.80 | 0.0014 | 0.01 |
| 390 | 12.09 | 0.0042 | 0.05 |
| 400 | 14.71 | 0.0143 | 0.21 |
| 410 | 17.68 | 0.0435 | 0.77 |
| 420 | 20.99 | 0.1344 | 2.82 |
| 430 | 24.67 | 0.2839 | 7.00 |
| 440 | 28.70 | 0.3483 | 10.00 |
| 450 | 33.09 | 0.3362 | 11.12 |
| 460 | 37.81 | 0.2908 | 11.00 |
| 470 | 42.87 | 0.1954 | 8.38 |
| 480 | 48.24 | 0.0956 | 4.61 |
| 490 | 53.91 | 0.0320 | 1.72 |
| 500 | 59.86 | 0.0049 | 0.29 |
| 510 | 66.06 | 0.0093 | 0.61 |
| 520 | 72.50 | 0.0633 | 4.59 |
| 530 | 79.13 | 0.1655 | 13.10 |
| 540 | 85.95 | 0.2904 | 24.96 |
| 550 | 92.91 | 0.4334 | 40.27 |
| 560 | 100.00 | 0.5945 | 59.45 |
| 570 | 107.18 | 0.7621 | 81.68 |
| 580 | 114.44 | 0.9163 | 104.86 |
| 590 | 121.73 | 1.0263 | 124.93 |
| 600 | 129.04 | 1.0622 | 137.07 |
| 610 | 136.35 | 1.0026 | 136.70 |
| 620 | 143.62 | 0.8544 | 122.71 |
| 630 | 150.84 | 0.6424 | 96.90 |
| 640 | 157.98 | 0.4479 | 70.76 |
| 650 | 165.03 | 0.2835 | 46.79 |
| 660 | 171.96 | 0.1649 | 28.36 |
| 670 | 178.77 | 0.0874 | 15.62 |
| 680 | 185.43 | 0.0468 | 8.68 |
| 690 | 191.93 | 0.0227 | 4.36 |
| 700 | 198.26 | 0.0114 | 2.26 |
| 710 | 204.41 | 0.0058 | 1.18 |
| 720 | 210.36 | 0.0029 | 0.61 |
| 730 | 216.12 | 0.0014 | 0.30 |
| 740 | 221.67 | 0.0007 | 0.16 |
| 750 | 227.00 | 0.0003 | 0.07 |
| 760 | 232.12 | 0.0002 | 0.05 |
| 770 | 237.01 | 0.0001 | 0.02 |
| | | | Sum = 1185.03 |

and these are multiplied by the appropriate weighting function at the same intervals and then summed. An example of such a calculation is given in Table 3.1.

### 3.1.6. Correlated colour temperature

The colour temperature of a source is the temperature of the black body or Planckian radiator having the same colour appearance or chromaticity as the source. For sources whose chromaticity does not exactly match that of a black body, a related quantity, the correlated colour temperature, can be defined. This is the colour temperature of the black body with the closest chromaticity to that of the source being considered, when the values are plotted on the CIE (1960) uniform-chromaticity diagram.

Because correlated colour temperature depends only on colour appearance it follows that light sources of widely differing spectral power distributions (e.g. incandescent lamps, fluorescent lamps, discharge lamps) can all have the same correlated colour temperature, and in practice correlated colour temperature is widely used as an index of colour appearance for all these types of lamp (Fig. 3.7).

Correlated colour temperature is also widely used to describe the colour appearance of sources other than lamps and notably the various phases of daylight, possibly the most common of which is $D_{65}$ (that phase having a correlated colour temperature of 6500 K). The CIE publishes standard data for illuminant $D_{65}$ (CIE, 1971) and this is shown in graphical form in Fig. 3.8 together with the spectral distribution of a practical $D_{65}$ simulator (a tungsten halogen lamp and filter) and a black body at 6500 K. It is clear from this graph that the simulator is very deficient in the ultraviolet, and this would lead to large errors if it were used to make measurements on fluorescent materials, for example. For all non-Planckian radiators, the spectral characteristics of the source should be quoted, as well as the correlated colour temperature.

Correlated colour temperature is a particularly useful concept in the case of tungsten filament lamps. These have a spectral power distribution in the visible which is very close to that of a black body, so that specifying the correlated colour temperature also defines the spectral power distribution. This makes them suitable for use as standards for quality control and specification purposes in the manufacture of signal lights and indicators, for instance, where the appearance of the coloured filters used depends critically on the spectral composition of the source used to illuminate them. A tungsten filament lamp at a stipulated correlated colour temperature provides a convenient source of known spectral power distribution.

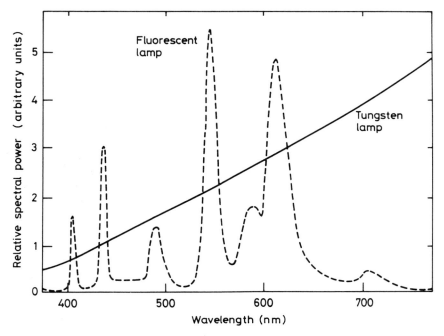

**Fig. 3.7**  Both these sources have a correlated colour temperature of 3000 K, but their spectral power distributions are very different.

The correlated colour temperature of a source is frequently determined by the use of a colorimeter. This consists of three detectors each with its own correction filter, which match respectively the $\bar{x}$, $\bar{y}$ and $\bar{z}$ $(\lambda)$ functions (i.e. the spectral tristimulus functions which define the CIE standard colorimetric observer). A diffuser is incorporated in front of the detectors such that they are uniformly illuminated. As in the case of photometers, the closeness of fit to the desired functions can be a limiting factor and for this reason colorimeter systems are not generally suitable for the direct measurement of non-Planckian radiators. For such sources, the correlated colour temperature is best determined by measuring the spectral power distribution (see Section 3.1.5) and calculating the chromaticity coordinates.

Spectral mismatch of the detectors can also lead to errors in the measurement of tungsten sources and for work of the highest accuracy the instrument should be calibrated with a colour temperature standard lamp. In addition, errors can arise if the lamp is placed too close to the colorimeter, resulting in non-uniform illumination of the detectors, or if there is inadequate screening for stray light. The latter can be a particular problem, since the use of a diffuser results in a very large field of view for

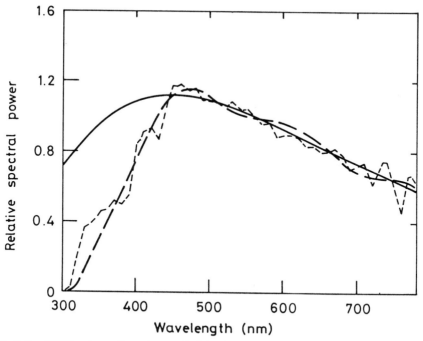

**Fig. 3.8**  CIE illuminant $D_{65}$ and a practical simulator. (------) Illuminant $D_{65}$; ( – – ) possible $D_{65}$ approximation; (——) black body at 6500 K.

the instrument. As in the case of photometric measurements, a small screen should be placed between the source and detector to check that the stray light screening is adequate.

## 3.2.  Calibration of Detectors

The previous sections have discussed the use of a calibrated source and a detector to measure other light sources. In such a situation, the detector is acting simply as a transfer device. In some cases, however, it is more useful to calibrate the detector itself and use it to make direct, rather than comparative, measurements on sources. The following sections will there-fore deal with the calibration of detectors and some of the possible sources of error in their use. Once again, the subject is dealt with from the users' point of view (i.e. an experimenter using a calibrated meter to measure sources) and will highlight the ways in which the method used for the calibration can affect the results of subsequent measurements.

### 3.2.1 Relative spectral responsivity

The equipment and techniques used in making spectral measurements on detectors bear many similarities to those used with sources, but in this case the system is calibrated with a reference detector, rather than a reference source. A typical arrangement is shown in Fig. 3.9. In this example, a monochromator is used to isolate a narrow wavelength band, so, as discussed in Section 3.1.5, the stray light properties should be checked and second order filters should be inserted as appropriate.

The monochromatic radiation is directed onto the reference and test detectors in turn, usually in a time-symmetric sequence to allow for source drifts, and the ratio between the outputs at each wavelength is equal to the ratio of the responsivities. The monochromator bandwidth should be as small as possible, whilst still providing acceptable signal levels from the reference and test detectors. For devices with spectral characteristics which are flat or vary only slowly with wavelength (such as the thermopiles and unfiltered photodiodes often used as reference detectors), the bandwidth of the incident radiation is not critical and the average responsivity for bands of different widths centred on a given value will be approximately constant. However many detectors designed for specialized applications (such as photometers and colorimeters) incorporate filters which isolate a particular portion of the spectrum or match a given action spectrum and for these, since the responsivity curve often shows sharp changes in slope, the bandwidth can have a significant effect on the result obtained.

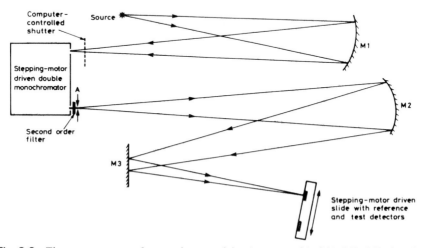

**Fig. 3.9** The measurement of spectral responsivity (not to scale). M1, M2, M3, imaging mirrors.

An imaging system is usually incorporated to focus either the exit slit of the monochromator or a fixed aperture onto the detector. It is important that this imaging should simulate the conditions under which the detector will be used as closely as possible. Many detectors show large spatial non-uniformities and for these the measured responsivity changes if the position or size of the irradiated area is changed. Any detector accessories, such as a fibre-optic pigtail attachment, can also change both the absolute and the spectral responsivity, so if these are to be used they should be included in the calibration.

Signals through monochromator systems are generally low and this often means a large difference between the calibration level and the levels under which the detector will be used. The linearity of the device should therefore be checked at several wavelengths; many detectors show appreciable differences in behaviour at different wavelengths.

A knowledge of the relative spectral responsivity characteristic allows corrections to be applied to measurements on sources of known spectral power distribution. Consider a detector of responsivity $s(\lambda)$ intended to represent an action spectrum $A(\lambda)$ and used to measure a source of spectral power distribution $S(\lambda)$. In order to allow for departures of $s(\lambda)$ from $A(\lambda)$, the following correction is required:

$$c = \frac{\int A(\lambda)S(\lambda)\,d\lambda}{\int s(\lambda)S(\lambda)\,d\lambda}.$$

This is analogous to the colour correction factor introduced in Section 3.1.1.

### 3.2.2. Absolute responsivity

The equipment used for determining relative spectral responsivity can also be used for making absolute responsivity measurements, provided the reference detector has been calibrated in absolute terms. The errors and precautions discussed previously apply equally in this situation.

It must be remembered that the result of a calibration of the absolute responsivity of a detector only applies at the wavelength at which the calibration is performed, although a measurement of the absolute responsivity at a given wavelength can be combined with relative responsivity data to give the absolute responsivity at all wavelengths of interest. Consider the case of a "black ray" meter, which is designed to measure the irradiance in the UVA region. Ideally it should have zero responsivity below 315 nm and above 400 nm, with constant responsivity between these wavelengths. No practical detector can match this exactly, but may have a characteristic

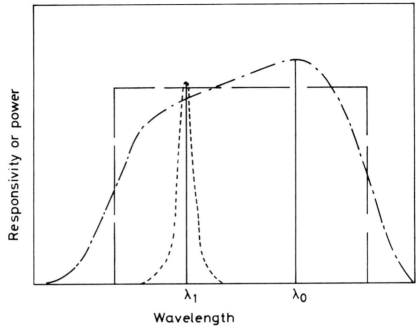

**Wavelength**

**Fig. 3.10** A calibration of the absolute responsivity at $\lambda_0$ will not apply for measurements on a source centred on $\lambda_1$. $(--)$ Desired responsivity; $(-\cdot-)$ actual responsivity; $(------)$ source spectral power.

similar to that shown in Fig. 3.10. Suppose this detector is calibrated at $\lambda_0$ and then used to make measurements on a source emitting in a narrow wavelength band centred on some other wavelength $\lambda_1$, say. If the ratio in the responsivities at $\lambda_0$ and $\lambda_1$ is not used to correct the results of the measurements, the values obtained will be seriously in error. Clearly, an absolute calibration at a single wavelength is of rather limited value. In general, relative values at other wavelengths are also required.

### 3.2.3.  Responsivity of thermopiles for total radiation

Thermopiles are designed to respond equally well to all wavelengths of optical radiation, although the degree to which they achieve this will depend on the quality of the black material with which they are coated and the transmission characteristics of any optical window which may be fitted. Such spectrally flat detectors are particularly suitable for determining the total radiant power emitted by a source, since they apply equal weight to all wavelengths.

Because thermopiles do not show a large change in responsivity with wavelength, and are in any case rather insensitive, they are usually calibrated using broadband radiation, rather than spectrally by comparison with a reference detector. Typically, one or more of the following sources are used: (a) a tungsten filament lamp run at a colour temperature of 2856 K, with a glass plate in front to absorb long wavelength radiation; (b) a similar lamp with a 20 mm thick water cell to absorb radiation of wavelengths above 1300 nm; (c) a nichrome radiator at a temperature of ~1270 K; (d) a black-painted surface at a temperature of ~470 K. A calibration with just one of these sources is sufficient if the source with which the thermopile is to be used is of similar spectral distribution. If this is not the case, then two or more can be used in order to assess the selectivity of the thermopile and to obtain an appropriate interpolated value for the responsivity.

Because the responsivity of a thermopile extends over such a wide wavelength range, special precautions are required in its use if background radiation is not to become a major source of error. All bodies emit thermal radiation according to Planck's radiation law (see Equation (10) in Chapter 1). For bodies at room temperature, most of this radiant energy is emitted at long infra-red wavelengths with insignificant amounts in the ultraviolet, visible and near infra-red. Thus it has no influence on the majority of detectors. But thermopiles are designed to respond to radiation at all wavelengths, including the far infra-red, and care has to be taken to ensure this does not introduce errors. The most satisfactory method for dealing with background radiation is to include it in the dark reading of the device, so that the increase in signal when the shutter is opened is due only to the source which is being measured. This means that the shutter must be at the same temperature as the surroundings, so that both the shutter and the surroundings radiate in the same way, and also that the temperature should not change in the course of the measurements. One way of achieving this is by the use of a water jacket or similar system to keep all parts of the equipment at the same, constant temperature. A possible arrangement is shown in Fig. 3.11. If a constant temperature enclosure is not available, the best solution is to keep all components sufficiently far from the source to prevent excessive heating and to avoid draughts which could cause the temperature to fluctuate.

Another common practice is to fix a glass filter to the front of the thermopile, which will absorb long wave radiation and effectively eliminate the effects of the background radiation. In this case, the filter must obviously be included in the actual calibration of the thermopile as well.

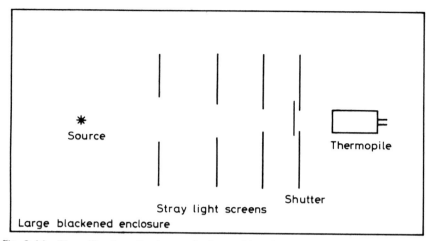

**Fig. 3.11** The calibration of a thermopile for total irradiance (not to scale).

## References

Commission Internationale de l'Eclairage (1971). *In* "Colorimetry", pp. 86–90. CIE Publication No. 15 (E-1.3.1), Bureau Central de la CIE, Paris.

Commission Internationale de l'Eclairage (1984). "The Spectroradiometric Measurement of Light Sources". CIE Publication No. 63, Bureau Central de la CIE, Paris.

Moore, J. R. (1980). "Sources of error in spectroradiometry", *Light. Res. Tech.* **12**, 213–220.

Saunders, C. L. (1972). "Accurate measurements of and corrections for nonlinearities in radiometers", *J. Nat. Bur. Stand.* **76A**, 437–453.

Walsh, J. W. T. (1965). *In* "Photometry", pp. 257–259. Dover, New York.

# 4

# Techniques for Spectroradiometry and Broadband Radiometry

## P. GIBSON

*Glen Spectra Limited*
*2–4 Wigton Gardens*
*Stanmore*
*Middlesex HA7 1BG, UK*

## B. L. DIFFEY

*Regional Medical Physics Department*
*Dryburn Hospital*
*Durham DH1 5TW, UK*

## 4.1. Comparison of Radiometric and Photometric Units with Biologically Effective Quantities

Radiometry is concerned with the measurement of optical radiation. Photobiologists are primarily concerned with the ultraviolet, visible and near infra-red regions (200–1000 nm) of the electromagnetic spectrum. Physically there are two characteristics of radiation which are of importance: wavelength and intensity. Spectroradiometry allows the determination of the intensity of the radiation from a source as a function of its wavelength, known as the spectral power distribution. The two radiometric measurements of most importance to the photobiologist are radiance from a source and irradiance at a surface, measured in $W.sr^{-1}.m^{-2}$ and $W.m^{-2}$, respectively. All biologically significant measurements can be related to radiometric measurements using the appropriate weighting function. Photometry is an attempt to quantify the sensation of the "brightness" response of the human eye to electromagnetic radiation between 380 and 780 nm. All photometric units are related to the unit of luminous flux, the lumen, which is the photometric equivalent of the watt (see Chapter 1).

The principle involved in converting from radiometric to photopic units can be applied to other biologically important quantities, such as photosynthetically active radiation and erythemally effective radiation. Figure 4.1

RADIATION MEASUREMENT IN PHOTOBIOLOGY
ISBN 0–12–215840–7

shows the ideal photosynthetic quantum response (Dodillet, 1961) and the reference erythema action spectrum recently proposed by the Commission Internationale de l'Éclairage (McKinlay and Diffey, 1987). In both cases, the biologically significant irradiance is the integral of the measured spectral irradiance, $E(\lambda)$, and the appropriate weighting function over the relevant

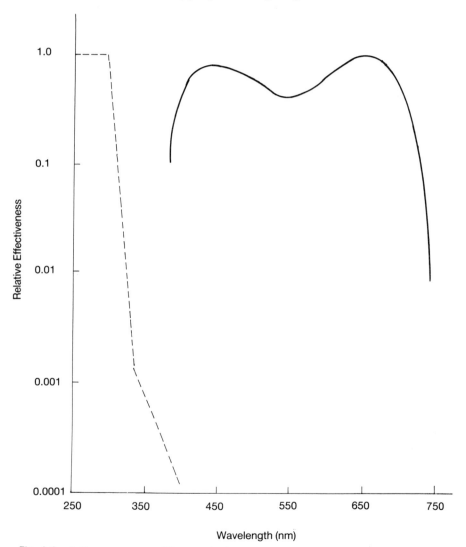

**Fig. 4.1** Action spectra. —— Photosynthesis (Dodillet, 1961); - - - skin erythema (McKinlay and Diffey, 1987).

wavelength range. Furthermore, as the two weighting functions do not overlap (Fig. 4.1), it is clear that it is meaningless to attempt to determine photosynthetically active radiation with a detector designed to measure erythemally effective radiation and vice versa. Similarly, as the weighting function used to convert spectral irradiance into photometric units is different from that for erythemally effective irradiance it is apparent that it is not valid to express erythemally effective irradiance in photometric units.

The measurement of the spectral power distribution of a source requires a spectroradiometer, whereas the measurement of luminous flux, photosynthetically active irradiance or erythemally effective irradiance can be made with a suitable broadband radiometer. Some of the fundamental aspects of spectroradiometry and broadband radiometry will be discussed below.

## 4.2. Spectroradiometers

### 4.2.1. Functional components

Spectroradiometers are instruments designed for the measurement of spectral radiance of sources and spectral irradiance from sources. A modern spectroradiometer consists of an input optic, a monochromator, an optical radiation detector, an amplifier, electronics for processing the output from the amplifier and for controlling the monochromator, and some form of data acquisition and display system.

The heart of the spectroradiometer is the monochromator, and because its spectral transmission characteristics depend upon the angular distribution and polarization of the incident radiation it is usual to use an input optic when measuring extended sources, such as low-pressure discharge tubes or daylight. To take account of the angular distribution of the incident radiation it is usual to use a cosine-weighted receptor. Two common cosine receptors are the integrating sphere and the dome diffuser, both of which are designed to collect radiation from $2\pi$ steradians. An integrating sphere is constructed in such a way that radiation does not pass directly from the input port to the exit port, but has to be reflected many times around the inside of the sphere. The larger the diameter of the sphere, the more reflections are required before radiation passes through the exit port into the monochromator. Barium sulphate and Teflon powder are used as coatings for the internal surfaces of integrating spheres used in the ultraviolet to near infra-red spectral regions. Neither of these materials are totally reflective, although both have reflectivities greater than 0.90, but less than unity, in the 250–1000 nm wavelength region, and therefore the larger the sphere the greater the attenuation.

Typically a 10 cm diameter sphere coated with barium sulphate is approximately 0.1% efficient in the 300–1000 nm region. Further details of integrating sphere performance are given in an excellent booklet published by Labsphere Inc. (Lovell, 1981). The advantages of the integrating sphere over a dome diffuser are that the cosine response is much closer to the ideal, and it can cover a wider wavelength range.

Dome diffusers are manufactured from a variety of materials, including ground quartz, Teflon, acrylic and opal glass. Ground quartz is used from 200 to 250 nm; beyond 250 nm it ceases to be a good diffuser, becoming much more transparent and therefore deviating significantly from the ideal cosine response. Teflon is usually used for dome diffusers in the 250–800 nm region, although one manufacturer recommends its use up to 1100 nm. Typically such diffusers have efficiencies of about 1%. Alternatively plane diffusers, such as a barium sulphate reflecting plaque, may be used; the efficiency of such a device is of the order of 10%.

Both the integrating sphere and the dome diffuser are highly efficient depolarizers of the incident radiation. This is an important consideration when measuring daylight, where the degree of polarization varies with the sky conditions.

Spectral radiance measurements of sources are made using imaging optics, either a telescope or microscope with a reflex viewer or a reflective mirror system. It is important to ensure that the aperture of the input optic is matched to the aperture of the monochromator, and in the case of refractive optics, that the optics transmit over the required wavelength region.

A classical monochromator consists of entrance and exit apertures, a collimating and a focusing mirror and a dispersing element, usually a diffraction grating. Before diffraction gratings became widely available, prisms were used. One major advantage of a grating monochromator over a prism monochromator is that for equal resolving power a grating monochromator has a very much larger transmitted spectral radiant power than a prism monochromator. One practical disadvantage of grating monochromators compared with prism monochromators is the passing of higher spectral orders, i.e. when set to 600 nm, a monochromator will pass second-order 300 nm and third-order 200 nm radiation. Therefore to cover an extended wavelength range with a grating monochromator it is necessary to incorporate an order-sorting device into the system. The order-sorting device usually comprises absorption filters.

There are two types of diffraction grating in use, the conventional ruled grating and the holographic grating which can have markedly different stray light and efficiency characteristics. Stray light from conventionally ruled gratings is in the form of "ghosts" and "grass", which are caused by periodic and random errors in the ruling process, respectively. Typically ghosts are of the order of $10^{-4}$ of the incident intensity whereas grass is of the order of $10^{-5}$. It is possible with care to manufacture holographic gratings without ghosts, and because of the interferometric nature of the manufacturing process grass is absent. There are however two other sources of stray light peculiar to holographic gratings, and although with extreme care it is possible to eliminate these sources it is often found that "grass" (especially that caused by dust in the photoresist) is typically of the order of $10^{-5}$. Often the other optical components in an instrument contribute significantly to the measurement of instrumental stray light. In general holographic gratings have better stray light characteristics than ruled gratings.

The efficiency characteristics of ruled gratings are generally superior to holographic gratings, except in the case of "blazed" holographic gratings. The mechanism of producing "blazed" holographic gratings is best suited to producing a blaze wavelength in the ultraviolet and such gratings can have an efficiency as high as that of an equivalent ruled grating. Typical grating efficiency curves are shown in Fig. 4.2. Holographic gratings can be made on a concave surface thus removing the need to have focusing mirrors in monochromators and thus reducing the losses inherent in monochromators using plane diffraction gratings. Concave holographic gratings do not have the wide range of ruling frequencies that are available from conventional

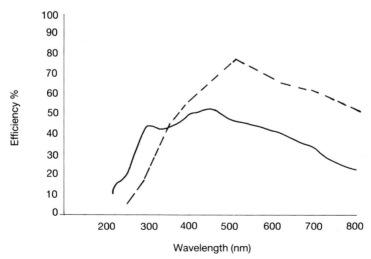

**Fig. 4.2** Typical efficiency curves of 1200 grooves per mm diffraction gratings; interferometrically produced with best efficiency in the visible (solid line), and classically ruled, blaze wavelength at 500 nm (broken line).

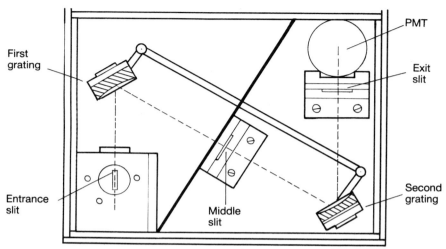

**Fig. 4.3** Optical layout of Optronic Laboratories Inc. Model 742 Spectroradiometer, a dual holographic grating monochromator. Reproduced with permission.

plane gratings and so tend to be used only in instruments for use in the 200 nm to 2.5 μm region. As concave gratings tend to be used to reduce the number of optical components and hence the cost of the instrument, concave gratings are not usually mounted in such a way as to allow the user easy interchangeability to extend the wavelength of his instrument. On the other hand, many instruments which use plane gratings offer the facility of kinematically mounted gratings for easy interchangeability.

In order to reduce the stray light further it is common to use double monochromators, where the exit slit of the first monochromator acts as the entrance slit of the second monochromator (Fig. 4.3). This arrangement reduces the stray light dramatically, and by doubling the dispersion at the exit slit improves the wavelength resolution available. Alternatively, the entrance and exit slit widths can be doubled to maintain the same bandwidth passing through the system, which more than compensates for the additional losses introduced by the second monochromator. The use of a double monochromator is especially important when measuring low levels of radiation, for example, solar radiation between 280 and 310 nm.

Instrument manufacturers offer both fixed and adjustable slits to define the bandpass of monochromators. On a practical note it should be stressed that fixed slits are preferred because they limit one potential source of error, incorrect slit width settings.

There are two detectors commonly used in spectroradiometers: photomultiplier tubes and silicon photodiodes. Although there are photomultipliers manufactured with S-1 responses which have been used at wavelengths as long as 1300 nm, most spectroradiometer manufacturers would use a photomultiplier only when working in the 200–800 nm region. The natural choice for work in the 300–1100 nm region is the silicon detector. The response of the silicon photodiode can be extended down to about 250 nm although the long wavelength responsivity tends to fall off somewhat as a result. Typically responsivity of the photomultiplier is of the order of $10^3$ times that of the silicon photodiode. The only advantages that the photomultiplier tube can offer over the silicon photodiode are its inherent sensitivity and its short wavelength responsivity. The disadvantages are the need for a high voltage power supply, relatively higher costs, fragility and spectral and spatial instability. Both detectors typically offer some six or seven orders of magnitude over which their output is linear with respect to the light intensity falling upon them. Thermal detectors, such as the pyroelectric detector (Section 2.3.3), can also be used, but these respond to changes in the radiation falling upon them and so require a mechanical chopper to modulate the incident radiation.

The choice of amplifier is determined by the type of detector chosen. If the detector is either a silicon photodiode or a photomultiplier the detector amplifier must be a low-noise trans-impedance type to convert the detector output current into a voltage. To handle the dynamic range of the detectors this amplifier must either be auto-ranging or have computer-controlled ranging. Such amplifiers tend to be slow to settle when dealing with rapidly changing signals. One alternative to such an amplifier is the logarithmic amplifier whose output is proportional to the logarithm to base ten of the input current. The resultant output is read by a panel meter to provide a front panel display and processed by an analogue to digital converter to provide digital information for subsequent data reduction by a personal computer. If a thermal detector is used it is necessary to have a synchronous amplifier, also known as a lock-in or a phase-sensitive amplifier. The low-noise trans-impedance amplifier, although relatively inexpensive, is susceptible to many sources of noise and interference. Although most amplifiers of this type have zero offset controls to compensate for "background" they cannot offset the AC component of the noise. The synchronous amplifier, on the other hand, can discriminate between the modulated signal and the AC component of the noise in addition to being able to remove a constant background. The resultant output is treated in the same way as the trans-impedance amplifier, again with either auto-ranging or computer-controlled ranging. Such amplifiers tend to be even slower to settle than trans-impedance types, and so do not find wide applicability except for low-level radiation sources, or for use with thermal detectors.

Many modern lock-in amplifiers have standard computer interfaces to facilitate data transfer to personal computers. The personal computer will control the wavelength drive of the monochromator, and even the order-sorting device. The advent of personal computers has removed the tedious data reduction process from the operator, and allowed the operator to concentrate on reducing potential sources of error in the measurements.

### 4.2.2. Calibration of spectroradiometers

Commercially available, or for that matter, home-made spectroradiometers can be calibrated using standard lamps available from national standards laboratories (see Chapter 3). To cover the spectral range from 200 to 2500 nm it is necessary to have two standard lamps: a deuterium arc lamp which has a continuum output which decreases from 200 to 400 nm and a tungsten halogen lamp which approximates to a black body source with a colour temperature of about 2900 K. These calibrated lamps when purchased directly from a national standards laboratory may be only two or three transfers from the national standard, but if purchased from a commercial source they may be between four and six transfers away from a national standard. Such calibrated lamps are usually supplied with instructions for use and suggestions for reducing transfer errors. It is important to ensure that the lamp is run at the same current and voltage as when it was calibrated, especially in the case of tungsten lamps which typically have uncertainties of the order of 3% of total irradiance for a 1% inaccuracy in current setting. It is important to calibrate the wavelength of the mono-chromator over the region of interest. A series of low-pressure discharge lamps emit spectral lines which can be used for the calibration of wave-length accuracy, e.g. low-pressure mercury discharge lamps emit prominent lines at 253.7, 435.8 and 546.1 nm. Other lamps which offer discrete lines are argon, krypton, neon and xenon. This procedure is especially important when working with a tungsten halogen calibration lamp and when measur-ing the solar spectrum in the 290–310 nm region where the spectral irradiance increases from around $10^{-5}$ to $10^{-1}$ W.m$^{-2}$.nm$^{-1}$. Table 4.1 indicates the change in spectral irradiance at various wavelengths from a tungsten halogen lamp for a 1 nm change in wavelength.

Simple guidelines have been proposed for consideration when calibrating spectroradiometers (Schneider and Goebel, 1981; Goebel and Schneider, 1981):

(1) Use a standard of spectral irradiance or radiance to calibrate a system for measuring spectral irradiance or radiance, respectively.

**Table 4.1** The change in spectral irradiance from a tungsten halogen lamp at various wavelengths for a 1 nm change in wavelength

| Wavelength (nm) | $\Delta E_\lambda$ per nm (%) |
|---|---|
| 250 | 7.5 |
| 270 | 6.0 |
| 300 | 4.0 |
| 350 | 2.5 |
| 400 | 2.0 |
| 500 | 0.88 |
| 700 | 0.25 |
| 1000 | 0.04 |

(2) Whenever possible select a calibration source similar to the test source in intensity and optical characteristics.

Unfortunately the second guideline is almost impossible to follow.

### 4.2.3. Sources of error in spectroradiometry

Accurate spectroradiometry, even where only relative spectral power distributions are required, requires careful attention to detail. Factors which can affect accuracy include wavelength calibration, bandwidth, stray radiation, polarization, angular dependence, linearity and calibration sources (Saunders and Kostkowski, 1978; Landry and Anderson, 1982).

High wavelength accuracy is particularly important when measuring the UVB component (280–315 nm) of terrestrial sunlight or solar simulators, since spectral irradiance changes rapidly with wavelength in the UVB region: 3% per 0.1 nm at 305 nm and 10% per 0.1 nm at 295 nm. Most spectroradiometers have a wavelength uncertainty of not better than 0.1 nm and a temperature coefficient of 0.1 nm.$°C^{-1}$ or higher.

The bandwidth and stray radiation characteristics of the monochromator are important because these result in output signals at wavelengths for which there is no radiation. The effect of bandwidth can be compensated for by correcting the observed spectrum by the slit function of the monochromator using a mathematical technique known as deconvolution (Shumaker, 1979). The effect of stray radiation can be minimized by using good quality optical components, particularly the grating, and preferably employing a double monochromator.

Insensitivity to polarization and a cosine-weighted angular response can be realized by using suitable input optics; either an integrating sphere or a properly shaped diffuser.

Since the spectral power distribution of sunlight and solar simulators in the UVB covers at least five orders of magnitude, the spectroradiometer must have a responsiveness factor whose variation with spectral flux is known, and preferably constant, over this entire range. This response should be linear to not less than a few per cent.

## 4.3.  Broadband Radiometers

### 4.3.1.  Functional components

A broadband radiometer is an instrument designed to measure the irradiance in a broad spectral band. Like a spectroradiometer such a broadband radiometer consists of an input optic, some form of wavelength selector, a detector and an electronics package. In general the input optic approximates to the dome diffuser described above for spectroradiometers, although more often in an attempt to make the sensor head more rugged the diffuser is flat or under a flat transparent cover. The wavelength selector is usually a stack of filters, either a combination of colour glass absorption filters or an interference filter with some additional blocking filters. The choice of detectors is usually increased to include vacuum phototubes along with silicon photodiodes and photomultiplier tubes. Vacuum phototubes have a distinct advantage over silicon photodiodes in that they can be made with a caesium–telluride photocathode which responds principally to wavelengths between 160 and 320 nm, thus leading to the term "solar-blind".

Although "solar-blind" photomultipliers are available they typically require a high voltage in excess of 400 V. The maximum anode supply voltage for a vacuum phototube is 100 V and the recommended supply voltage is only 15 V, easily within the capability of battery operation in portable equipment. Two other detectors are available for use in broadband radiometers: selenium photocells and gallium arsenide phosphide (GaAsP) photodiodes. Neither offer the wide wavelength coverage, nor the high radiant sensitivity of silicon. However, GaAsP photodiodes have no spectral response beyond 700 nm, and when combined with a "black glass" (e.g. Schott UGI) filter, result in an ultraviolet-selective detector (Wilson and Lyall, 1986). Selenium suffers from "fatigue", that is its output for a given irradiance level decreases over a period of several years and while this will not be a problem in a laboratory with a periodic calibration regime, it is an undesirable feature.

By careful selection of these three components, diffuser, filter stack and detector, the required spectral response can be obtained. Great care should be taken with the selection of the filter stack when using a silicon photodiode as detector to ensure that the transmission out of the required band is minimal as the sensitivity of the detector may be some five times greater to the out-of-band radiation than the in-band radiation. In an extreme situation, if the ratio of the intensity of the out-of-band to the in-band radiation is 1000:1, the ratio of the transmittance of the filter stack of the out-of-band to the in-band radiation is 1:1000 and the ratio of the detector sensitivity to out-of-band to in-band radiation is 5:1, then the output of the radiometer would be in error by a factor of 5.

The electronic circuitry to convert the current from the radiation detector into a reading for the operator is once more a trans-impedance amplifier. Most modern battery-operated instruments will utilize C-MOS integrated circuits to ensure low power consumption, to extend the mean time between failures and to reduce sources of noise within the electronics. Electronics have progressed over the last decade to such an extent that 10 years ago a simple moving coil meter was the standard readout device, whereas nowadays it is possible to have a portable datalogger capable of capturing some 3000 readings to be read directly into a computer at a later time.

### 4.3.2. Physical performance

The physical performance of broadband radiometers is governed by the spectral sensitivity, the angular response and the linearity of response of the sensor. By careful selection of the components in the sensor assembly almost any desired spectral response can be achieved. In general, commercial broadband radiometers are limited by the spectral response of silicon. In some cases it is desirable to have a matched spectral response, i.e. the photopic response, but in other cases the spectral response of a sensor is accepted as the best attainable over a particular spectral region.

The naive users of broadband radiometers often gain the impression from commercial literature that instruments are readily available to measure UVA, UVB or UVC. In order to meet the criterion for a UVB radiometer, say, the sensor should have a uniform spectral response from 280 to 315 nm (the UVB waveband) with zero response outside this interval. In other words, the electrical output from the sensor should depend only on the total power within the UVB waveband received by the sensor and not on how the power is distributed with respect to wavelength. In practice no such sensor exists with this ideal spectral response (neither does one exist that measures UVA or UVC correctly for that matter). All radiometers which combine a photodetector with an optical filter have a non-uniform spectral sensitivity within their nominal spectral band. This results in a displayed irradiance which depends not only on the true broadband irradiance, but also on the spectral power distribution of the source. This problem has been discussed in more detail elsewhere (Diffey, 1982; Mackenzie, 1984; Levin, 1986).

Even in those cases where the spectral response of the available broadband radiometer does not match the desired action spectrum it is possible to make meaningful measurements provided there is a degree of overlap between the spectral response of the sensor and the action spectrum. A simple method for relating the erythemally effective irradiance and measured UVB irradiance has been described by Diffey (1986) which could be extended to any other two parameters related in a similar way.

Most sensors claim to offer a good cosine response to the incident radiation, since many photobiological media are said to exhibit a cosine response. In some cases where a well-collimated beam of radiation is used to produce a well-defined exposure area, the incident radiation would be at or near normal incidence to the sensor. In these cases a simple, filtered detector would be adequate. In other cases, where extended sources are used, a sensor with a cosine or Lambertian spatial response is required. The closeness of perfect cosine response to that attained by a sensor can be displayed in several ways. Two useful presentations are either to plot the percentage deviation from the cosine response versus the angle of the incident ray, or to plot the percentage deviation of the half-angle subtended by the source at the sensor versus the half-angle subtended. Two less informative plots, often provided by manufacturers, are to plot the actual sensitivity versus the angle in polar or linear coordinates. The representation for a perfect cosine receptor is circular or a cosine function in polar or linear coordinates, respectively. In both cases, deviations of the measured response from the ideal response are difficult to assess.

The linearity of the radiometers' response as a function of the intensity of radiation falling upon its sensor is an important parameter. In many cases it is only necessary to establish linearity over two or three orders of magnitude, but in some cases radiometers covering five orders of magnitude are available. Theoretically, the radiometer response should be linear with irradiance. In practice, as the irradiance increases, the radiometer response will tend to become lower than expected; this deviation depends upon the internal characteristics and the load resistance of the detector, becoming less marked at lower load resistance. Typically linearity of about $\pm 1\%$ is achieved through use of trans-impedance amplifiers. Some manufacturers quote figures of $\pm 1\%$ at irradiance levels some 10 times greater than the maximum their equipment is designed to measure.

### 4.3.3. Calibration

The calibration of broadband radiometers is probably one of the most difficult tasks to be undertaken. One method commonly used by national standards laboratories and many manufacturers is the use of electrically calibrated pyroelectric radiometers (ECPR). When chopped radiation and gated electrical power are applied out of phase to a pyroelectric device, lock-in detection of the output signal will yield a difference signal. By adjusting the electrical power to produce a null signal it is possible to determine absolute radiant power. Methods using standard tungsten and deuterium lamps are also employed, but the use of standard lamps and ECPRs is likely to include significant out-of-band components unless precautions are taken.

Ideally, broadband radiometers need to be calibrated separately for each type of light source for which they are to be used. The method of choice is as follows.

(i) Measure the spectral irradiance from the desired light source using a spectroradiometer.

(ii) From the spectral irradiance data, determine the irradiance within the waveband of interest (either unweighted or spectrally weighted according to some pre-defined action spectrum).

(iii) Place the entrance aperture of the broadband radiometer at the same point in space as the entrance optics of the spectroradiometer and note the reading.

(iv) Either adjust the gain control of the radiometer electronics to read the correct irradiance or determine a multiplicative correction factor by dividing the correct irradiance by the displayed irradiance. If the radiometer is intended to be used with more than one type of lamp the latter method is preferred.

Finally, it should be remembered that the sensitivity of all radiometers will change with time—frequent exposure to high intensity sources of light will accelerate this change. For this reason it is always a sound policy to acquire two radiometers, preferably of the same type, one of which has a calibration traceable to a national standards laboratory. This radiometer should be reserved solely for intercomparisons with the other radiometer(s) used for routine purposes. A measurement of the same source is made with each radiometer and a ratio calculated. It is the stability of this ratio over a period of months and years which indicates long-term stability and good precision.

## References

Diffey, B. L. (1982). "Ultraviolet Radiation in Medicine", pp. 77–83. Adam Hilger, Bristol.

Diffey, B. L. (1986). "Possible Errors Involved in the Dosimetry of Solar UV-B Radiation". *In* "Stratospheric Ozone Reduction, Solar Ultraviolet Radiation and Plant Life" (R. C. Worrest & M. M. Caldwell, Eds), pp. 75–86 Springer-Verlag, Berlin.

Dodillet, H. J. (1961). Der Maximalwert des phyto-photometrischen Strahlungs-äquivalentes. *Lichttechnik* **13**, S556–558.

Goebel, D. G. & Schneider, W. E. (1981). Automatic Systems for Spectroradiometric Measurements. SPIE Vol. 262, Light Measurement '81, pp. 94–104.

Landry, R. J. & Anderson, F. A. (1982). Optical radiation measurements: instrumentation and sources of error. *JNCI* **69**, 155–161.

Levin, R. E. (1986). Radiometry in photobiology *In* "The Biological Effects of UVA Radiation" (F. Urbach & R. W. Gange, Eds), pp. 30–41. Praeger, New York.

Lovell, D. J. (1981). Integrating Sphere Performance. Labsphere Inc., North Sutton, New Hampshire 03260, USA.

Mackenzie, L. A. (1984). UV radiometry in dermatology. *Photodermatol.* **2**, 86–94.

McKinlay, A. F. & Diffey, B. L. (1987). A reference action spectrum for ultraviolet induced erythema in human skin. *CIE J.* **6**, 17–22.

Saunders, R. D. & Kostkowski, H. J. (1978). Accurate Solar Spectroradiometry in the UV-B. Optical Radiation News No. 24, US Department of Commerce, NBS, Washington.

Schneider, W. E. & Goebel, D. G. (1981). Radiometric Standards and Sources for Calibration of Optical Radiation Measurement Systems. SPIE Vol. 262, Light Measurement '81, pp. 74–83.

Shumaker, J. B. (1979). *In* "Self-study Manual on Optical Radiation Measurements: Part 1—Concepts" (F. E. Nicodemus, Ed.), pp. 35–90. NBS Technical Note 910-4, US Department of Commerce, Washington.

Wilson, A. D. & Lyall, H. (1986). Design of an ultraviolet radiometer. 2: Detector optical characteristics. *Appl. Opt.* **25**, 4540–4546.

# 5
# Action Spectroscopy

M. G. HOLMES

*Department of Botany*
*University of Cambridge*
*Downing Street*
*Cambridge CB2 3EA, UK*

## 5.1. Introduction

Action spectroscopy is used to determine the spectral characteristics of "functional" pigments. Action spectra have been used for many years to identify the functional pigment, or photoreceptor, of a response by comparing the absorption spectrum of the putative photoreceptor to the action spectrum of the response. As an analytical procedure, action spectroscopy has several advantages. Two of the most important are that it is non-destructive and that the transfer of the active factor (i.e. radiation) to the target (photoreceptor) is immediate.

There are also many potential pitfalls and problems to be overcome in action spectroscopy. Many of these concern the methodology of experimental procedure and data analysis. Strict rules must be followed in terms of understanding response linearity, temporal changes in the response, reciprocity, etc., and are not within the scope of this chapter. However, the most difficult problems are usually the many aspects of radiation measurement which are critical for accurate action spectroscopy. These can be divided into two main groups.

The first group derives mainly from practical problems of radiation attenuation by the target tissue. The optical properties of the target material can distort action spectra because the radiation measured as arriving at the target does not necessarily represent the radiation absorbed by the photoreceptor. This introduces error because the Grotthus–Draper Law (the first law of photochemistry) states that only absorbed radiation can cause photochemical reactions. The main reasons for optical properties of the target material distorting action spectra are the existence of energy flux gradients within the tissue, and the sieve effect which results from heterogeneous distribution of the absorbing materials. These factors will be discussed in detail in Chapter 9.

RADIATION MEASUREMENT IN PHOTOBIOLOGY
ISBN 0–12–215840–7

The second group of problems with radiation measurement in action spectroscopy is the wide range of factors which must be accounted for. Many of these are common to other aspects of photobiology, but some are specific to action spectroscopy and are summarized below.

Action spectroscopy involves the use of artificial monochromatic sources. In most studies, this means that the detector must have a wide range of spectral sensitivity and that very accurate and precise measurements of wavelength must be made to separate action maxima and minima. If the spectral quality of that radiation is known, a simple detector with a calibrated broad-waveband response may be adequate. It is important in this type of measurement that there are no spectral impurities and that the sensitivity of the detector at each wavelength is known. If spectroradiometric measurements must be made, then careful attention has to be paid to the relationship between the instrument's dispersion system and the bandwidth of the monochromatic sources.

Action spectroscopy can demand very accurate and precise measurements of fluence rate to separate dose response curves. In addition, it may be necessary to measure a very wide range of fluence rates. Both of these requirements can result in problems in choosing the correct detector.

One of the most difficult aspects of radiation measurement in action spectroscopy is that of choosing the correct receiver geometry. This is particularly important for quantitative comparisons of action spectra and for spectral identification of the photoreceptor.

## 5.2. Monochromatic Sources

The monochromatic sources used in action spectroscopy present specific problems in radiation measurement. The most frequently overlooked aspect is that manufacturers' specifications for filters are not always reliable. One reason for this is that filters can deteriorate with time and with careless treatment. Pigmented glass or plastic can be modified by high fluence rates and by heat, and the transmission characteristics of interference filters are changed by high temperature and by the ingress of water.

Another reason for not relying on manufacturers' data is that they are often inaccurate. It is important that there is precise knowledge of the blocking characteristics of the filter combinations used. For example, ultraviolet and blue-transmitting filters often have a second transmission peak in the far-red. Secondary peaks can be in a waveband of high sensitivity of the pigment and therefore distort grossly the true absorption spectrum. Many filters are not symmetrical in their transmission, so it is preferable to determine the central wavelength rather than the peak wavelength or wavelength maximum (Fig. 5.1). This is particularly important when it is recognized that many interference filters have broad, or even double, wavelength maxima.

Whereas attention to the above points will minimize error in radiation

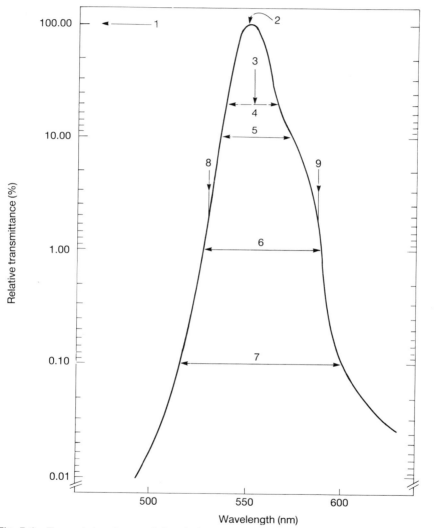

**Fig. 5.1** Transmission characteristics of a bandpass filter. 1 = peak transmission or transmission maximum; 2 = peak wavelength or wavelength maximum; 3 = central wavelength; 4 = half bandwidth, half-power bandwidth, or bandwidth; 5 = 1/10 power bandwidth; 6 = 1/100 power bandwidth; 7 = 1/1000 power bandwidth; 8 = cut-on (5%); 9 = cutoff (5%).

measurement in action spectroscopy, selective attenuation of radiation by the target tissue may still pose difficulties. Some approaches for overcoming these are given in Chapter 9. One of the most troublesome, however, can be fluorescence caused by the actinic source. In chlorophyll-containing tissue, for example, pure blue sources can induce a red fluorescence which may cause significant activation of red-absorbing photoreceptors, thereby leading to erroneous interpretation of the biological response.

## 5.3.  Detector Types

Discussion of detector types is restricted to a brief overview of the suitability of various detector types in action spectroscopy. Greater detail can be found in Chapter 2. There are two main types of detector, these being the thermal radiation detectors and the photoelectric or quantum detectors. The thermal radiation detectors used in action spectroscopy are of the thermoelectric type and convert heat energy to electrical energy. The photoelectric detectors are of various types, and it is important that the correct ones are chosen for action spectroscopy studies. The requirements of action spectroscopy for choosing the appropriate detector are great, especially in the plant sciences, because the photoreceptors studied usually have a wide range of spectral sensitivity and are sensitive to changes in energy flux over several orders of magnitude.

### 5.3.1.  Thermoelectric detectors

Thermoelectric detectors are extremely useful in action spectroscopy because they typically have a non-selective response to different wavelengths and have a very broad waveband of spectral sensitivity, and offer a fairly linear response to a wide range of radiation levels. The usual disadvantages of thermoelectric detectors are their relatively slow response to small and rapid changes in radiation, and the sensitivity to changes in the temperature of the experimental environment.

#### 5.3.1.1.  Thermopile

The thermopile is one of the most accurate instruments for radiation measurement and is capable of giving highly reproducible measurements. They require only periodic calibration if handled correctly. They are particularly suitable for the calibration of other radiation-measuring equipment, and for the measurement of radiation emitted by monochromators, lasers and filter monochromator systems. They are therefore ideal instruments for most action spectroscopy studies if the conditions for suitable receiver geometry are met.

The temperature sensitivity of thermopiles can cause both direct and indirect problems. The direct problems arise from failure to thermally adapt the instrument to the working environment, and failure to screen thermal sources which can influence the measurements. There are two common indirect problems. The first is the spectral change in sensitivity which can result from the addition of windows to prevent thermal effects of air circulation. Whereas thermopiles have an almost uniform response to the UV, visible and near-IR wavebands, the addition of a window will modify the absolute sensitivity and can influence the spectral sensitivity. Major problems are unlikely if a crystal quartz window is fitted because the transmittance is fairly uniform over the range from approx. 250 to 3000 nm.

The second common indirect problem is caused by the use of removable filters. As with the thermopile itself, the removable filters must be totally thermally adapted to the environment in which the instrument is used. If the filter is thermally adapted, it will cause no meter deflection when placed over the open aperture in darkness. If an IR-transmitting filter is used to correct for extraneous heat sources, it is important that the whole transmission spectrum is checked because several of these IR-transmitting filters also transmit in the UV.

Thermopiles are capable of measuring the highest energy fluxes which will be used in action spectroscopy, but prolonged periods of monitoring at very high fluxes should be avoided. Difficulties usually result when trying to measure very low levels of energy. Greater sensitivity and a steadier zero can often be obtained with vacuum cases, but this can result in an increased response time. At fluxes of less than about $10^{-2}$ W.m$^{-2}$ it will usually be necessary to use a photoelectric detector.

### 5.3.1.2. Bolometer

Although thermistor bolometers are only slightly less sensitive than thermopiles, they are also very sensitive to extraneous heat. The difficulty is that the complications involved in removing error from this source are greater than with thermopiles. The requirement for an accurately controlled external power supply and the difficulty in attaching spatial receivers for measuring diffuse radiation usually make bolometers less convenient than thermopiles as a thermoelectric detector.

### 5.3.2.  Photoelectric detectors

The most important point to bear in mind when considering photoelectric detectors for radiation measurement in action spectroscopy is that they have a strong wavelength dependency. It is therefore imperative that they are rigorously calibrated for all wavelengths. Having said that, photoelectric detectors are extremely sensitive to low radiation levels, are relatively insensitive to temperature changes, are rugged and usually provide good linearity. Photoelectric detectors can be divided into photoconductive, photoemissive and photovoltaic groupings.

#### 5.3.2.1.  Photoconductive detectors

Photoconductive detectors operate on the principle that the resistance of some semiconductors changes when irradiated and that this change can be used to measure the energy flux. Although their general characteristics are good, most are sensitive to the longer, infra-red wavelengths and are therefore not generally suitable for action spectroscopy studies (Section 2.4.4).

#### 5.3.2.2.  Photoemissive detectors

There are two main types of photoemissive detectors, these being the phototube and the photomultiplier. Both have relatively limited wavebands of spectral sensitivity, although different detectors covering a specific waveband of several hundred nanometers can be purchased and used (see Section 2.4.1). Photoemissive detectors excel where fast time constants are required or where very low levels of radiation have to be measured.

Phototubes consist of an anode and a cathode enclosed in a glass or quartz envelope. The spectral sensitivity depends mainly on the composition of the cathode. The current is usually measured via an external amplifier. Phototubes are inferior to photomultipliers in most radiation measurement applications for action spectroscopy because they suffer to various extents from relatively low sensitivity, high noise levels and non-linear responses at high fluxes when compared with photomultipliers.

Photomultipliers consist of a series of dynodes which amplify the signal within the tube. Used in conjunction with a thermopile, the combination provides all the requirements for radiation measurement in action spectro-scopy. Photomultipliers offer extremely high sensitivity for working at low radiation levels, but also give excellent linearity over a wide range. Very high fluxes can cause temporary error, especially at high operating voltages. Permanent damage is unlikely. If temporary damage (a change in sensitivity, known as fatigue) occurs, this can be rectified by a few minutes of darkness; in extreme instances, it may also be necessary to reduce the applied voltage and switch off the instrument for a few minutes. These detectors require particularly careful treatment. New tubes should be run for about 50 h before calibration and use to avoid the exaggerated ageing during this period. They should then be handled gently and calibrated at least twice yearly, even if not used in the interim period, because they do suffer from poor and unpredictable stability.

### 5.3.2.3.  Photovoltaic cells

Photovoltaic cells consist of a low internal resistance semiconductor layer sandwiched between a thin transparent metallic film (the cathode) and a solid metal base (the anode). The incident radiation generates an EMF which is measured by a microammeter, although this EMF is high enough for the instrument to be self-powered at high energy fluxes. Discussion is restricted here to the two main established and proven semiconductor materials used as photovoltaic cells which are selenium and silicon. A discussion of junction photodetectors (photodiodes) which are recent developments of photovoltaic cells is given in Section 2.4.3.

The selenium photocell is very suited to photometric instruments because the spectral response is very similar to the average luminous efficiency curve of the human eye. These photocells are useful in many animal photobiology studies which tend to have a limited window of this type, but can prove restrictive to the plant photobiologist who must consider broader wavebands. Some selenium photocells do, however, have a secondary peak of sensitivity in the 650–800 nm waveband. Overall, the selenium photocell does not compare favourably with the silicon photocell. Selenium photocells suffer from fatigue and non-linear responses at high energy fluxes, from relatively poor stability and from temperature-induced errors.

The silicon photocell is superior to the selenium photocell in its photocurrent sensitivity, its temperature sensitivity, its linearity, its susceptibility to fatigue and its compact size. The final choice may depend on wavelength sensitivity, with selenium cells covering the approximate 300–800 nm range and the silicon cells covering the approximate 400–900 nm range. As mentioned earlier, continual changes in the field of photon detector development need to be monitored when considering the purchase of a new instrument (see Chapter 2).

## 5.4. Linearity

The linearity of a detector's response is its ability to follow accurately gradual changes over a wide range (usually several decades) of radiation levels. In most cases, non-linearity results from changes in the sensitivity of the sensor, or from inaccurate range factor switching.

Various means can be used to measure linearity, the commonest being varying the energy flux by changing the source–sensor distance, varying the output of the source or using two radiation sources. Of these, the dual radiation source approach is the most reliable. To do this, two radiation sources, x and y, are aimed at the detector. Source x is used to provide various energy flux levels, and source y is used as a constant output source. The output of x is increased in stages from zero up to the maximum available and the measurements recorded alternately with y switched on and with y switched off. The signal difference will be constant if the instrument is perfectly linear in its response. An alternative approach with dual radiation sources is to use the addition technique. In this case, the signal from x + y together should be equal to the sum of x alone and y alone.

## 5.5. Measuring Low Radiation Levels

One of the commonest problems in action spectroscopy is the measurement of low radiation levels. If neutral density screens or filters are used to produce the low radiation levels, and their transmission characteristics are known, then it is possible to calculate the energy flux at values which are below those which the instrument can measure. However, it is often possible to make direct measurements at radiation levels approaching the limits of the equipment's sensitivity if certain basic precautions are taken.

In an ideal situation, the same instrument will be used for all radiation levels which have to be measured, but it has to be recognized that action spectroscopy involves accurate and precise measurements over several orders of magnitude of energy flux, and that a wide range of wavelengths may be used, thereby placing an increased demand on the detecting instrument. The first requirement is therefore that the detector is appropriate to the waveband to be measured. The next objectives are to maximize the signal input and to minimize the amount of noise.

The first step in maximizing the signal is to ensure that the amount of radiation reaching the receiver surface is not unnecessarily restricted. The largest possible receiver surface should be used to provide radiation for the photodetector. It is then important that the use of major attenuators between the receiver and photodetector be avoided. Fibre-optic radiation guides can severely reduce radiation input, especially if many of the fibres are broken. In this context, a minimum number of air/glass interfaces should be used as each results in unnecessary loss of radiation. If a monochromator is used, it is often possible to increase the slit width as long as the limits imposed by the type of radiation source which is to be measured are borne in mind.

Noise can usually be reduced by increasing the sampling time, so longer photometer response times can often be obtained by increasing the scanning time in spectroradiometric measurements. Thermionic emission of electrons from the photocathode or dynodes can cause a problem with photo-multipliers operating at very low radiation levels, but this can often be overcome by cooling the photomultiplier because the dark current increases with temperature. Another alternative is to increase the voltage applied to the photomultiplier in order to increase its sensitivity, but a stage is reached where the advantages of increased sensitivity are offset by the dispropor-tionately high increases in dark current at high voltages. Although a high-pitched whine at high applied voltages is a normal occurrence and can be ignored, operating the photomultiplier within about 250 V of the maximum rated voltage should be avoided.

## 5.6. Receiver Geometry

The choice of correct receiver geometry in action spectroscopy depends on three factors. These are the angle of incidence of the actinic radiation to the receiver, the spatial orientation of the irradiated target material and the contribution of reflected or scattered radiation arriving from outside the acceptance angle of the receiver. In the majority of action spectra studies, the radiation can be considered to be parallel beams from a point source, and the receiver geometry requirements appear to be straightforward. Difficulties usually arise when the receiver geometry is assessed with regard to the spatial characteristics of the target material and with regard to the analysis of stray radiation. The type of receiver used and the reflectivity of the surroundings have major effects on the radiation measurements.

As can be seen from Table 5.1, the acceptance angle affects markedly the radiation measurement, in particular when the surroundings of the target are reflective. It can be argued that the choice of incorrect receiver geometry would result in a systematic error which would not necessarily affect the shape of the resulting action spectrum. However, this approach would result in an inability to compare the data quantitatively with other measurements in which the correct receiver geometry had been used. More importantly, the choice of incorrect receiver with regard to the dimensions

**Table 5.1**  The influence of receiver type and background reflectivity on radiation measurement. The measured photon fluxes are expressed as a percentage of that recorded by the spherical detector when the lamp was against a white background[a]

| Receiver type | Black background | White background |
| --- | --- | --- |
| Normal incidence | 13 | 15 |
| Cosine corrected | 36 | 37 |
| Spherical | 53 | 100 |

[a] The radiation was produced by a horizontally positioned $1.1 \times 1.2$ m monochromatic red radiation source ($\lambda_{max} = 658$ nm) composed of Philips TL40W/15 fluorescent lamps filtered through a 3 mm thick sheet of No. 501 Plexiglas suspended 0.7 m above a black background (black cloth) or a white background ("Pearl White" formica). A comparison is made of radiation measurements with normal incidence response (acceptance angle $= 9.5°$), cosine response and spherical response receivers. (From Hartmann, 1977.)

of the target and the possible location of the photoreceptor would result in a much reduced ability to relate the experimental observations to the objective of action spectroscopy which is the spectral identification of the photoreceptor.

### 5.6.1.  Normal incidence detectors

Detectors with a very restricted acceptance angle ("normal incidence detectors") are only appropriate for measuring parallel bundles of radiation from a point source and where the incident radiation arrives within the nominal aperture angle. Even then, their use is correct only if there is negligible reflection or scattering at the plane of irradiation (Fig. 5.2). Normal incidence detectors are therefore only suitable for applications where the incident radiation arrives within the nominal aperture angle. This excludes their use in many experimental situations, such as natural daylight, radiation from fluorescent lamps and most growth cabinet lighting. A further important prerequisite for the use of normal incidence detectors is that the biological target should develop predominantly in two dimensions and that these dimensions should be in a plane at right angles to the incident radiation. If any of the above conditions cannot be met, then the use of a cosine-corrected or a spherical response receiver will be appropriate.

### 5.6.2.  Cosine-corrected receivers

A cosine-corrected receiver should be used with diffuse radiation sources if the angle of incidence is predominantly $< 180°$ (Fig. 5.2). This infers that

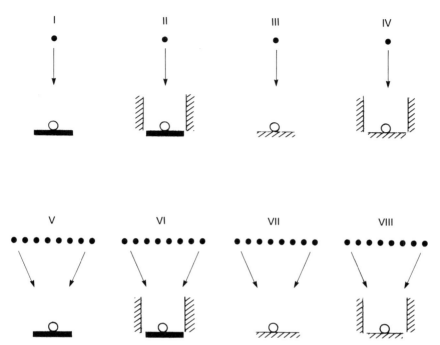

**Fig. 5.2**  Factors influencing the choice of correct receiver geometry. Detectors with a very restricted acceptance angle ("normal incidence") are only suitable for measuring point sources (I) in the absence of reflection or scattering. If the target material scatters this radiation and a significant amount can be reflected from the surroundings, then a cosine (II) or spherical response (III, IV) receiver will be required. With diffuse radiation sources, a cosine-corrected receiver is appropriate if the angle of incidence is $< 180°$ (V and VI). If a significant proportion of radiation arrives from outside this angle (VII and VIII), a spherical response receiver will be required. A point source is considered here as a radiation source whose cross section is less than 10% of the distance from the source to the irradiated object. Key: • = point source (e.g. projector), ••••• = diffuse source (e.g. fluorescent lamps), ▬ = non-reflective surface, ⁊⁊⁊ = reflective surface, ○ = irradiated object.

the radiation from the source arrives within this solid angle and that there is negligible scattering and reflection from outside this range. A cosine-corrected receiver can also be used with point sources if the radiation arrives at right angles to the receiver surface. As with normal incidence detectors, it is important that the target develops predominantly in two dimensions in a plane at right angles to the incident radiation.

The reason for using a cosine-corrected receiver is that the flux arriving at a flat surface decreases as a function of the cosine of the angle of incidence. It is therefore necessary to apply a proportionality factor to radiation which is arriving at other than normal incidence. The factor applied is $1/\cos\theta$

where $\theta$ is the angle between the plane of incidence of the radiation and the normal to the surface. If a receiver were perfect, the response at 45° off-axis would be the cosine of 45°, i.e. 0.707 of the response at normal incidence, and at 60°, 0.500 of the response, etc. Some examples of cosine-corrected receivers and their deviations from a true cosine response are shown in Fig. 5.3.

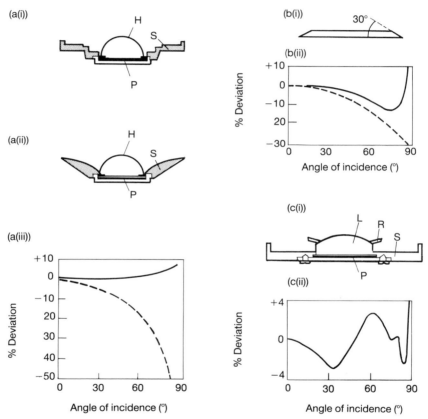

**Fig. 5.3** Examples of cosine-corrected receivers. (a(i) and a(ii)) Receivers corrected by metal screens. H = hemispherical opal glass bowl, P = photocell, S = light metal screen. (a(iii)) Deviation from the cosine law of a photocell without (broken line) and with (continuous line) the corrected receiver. (Redrawn after Eckhardt, 1965.) (b(i)) Opal glass plate cut to the shape of a cone frustrum. (b(ii)) Deviation from the cosine law of a photocell without (broken line) and with (continuous line) the correction plate. (Redrawn after Kubín, 1971.) (c(i)) Receiver corrected by a screen and a convex lens. L = lens, S = screen, R = ring attached to the lens, P = photocell. (c(ii)) Deviation from the cosine law of the combined photocell and correction system. (Redrawn after Dlugos, 1958.)

No receivers have perfect cosine correction and errors in the range of 5–10% deviation at off-normal angles of 80° or greater must be accepted. Such errors are of little significance unless the proportion of radiation arriving at these extreme angles is substantial in comparison with that arriving perpendicular to the receiver. If the photosensitive region of the irradiated tissue is known, and it lies predominantly at other than normal incidence to the actinic beam, then the base of the cosine-corrected receiver should be angled to lie parallel with the surface of the photosensitive region. The objective of this procedure is to compensate for the reduced incident flux per unit tissue area.

There are several methods by which photodetectors can be modified to provide a cosine-corrected response to the incidence radiation. Among the simplest methods is the placing of a sheet of white chromatography or filter paper, or cloudy white Perspex or Plexiglas over the photodetector. This will produce an acceptable cosine correction up to about 60° from normal incidence. A better alternative is the use of a flat opal glass, preferably with a rough rather than a smooth, polished surface. Slight improvements can be gained by altering the height of the glass relative to the photodetector, by using a screen to prevent radiation arriving at extreme angles, by shaping the glass plate into a cone frustrum, or by using a hemispherical globe of opal glass (Fig. 5.3).

### 5.6.3. Spherical response receivers

When the radiation arrives at a solid angle of > 180°, or the target material is predominantly three-dimensional, then a spherical response receiver should be used (Fig. 5.4). Spherical response receivers can be bought commercially for a limited range of radiation-measuring instruments, or they can be constructed with relative ease. When the radiation propagates predominantly in two opposite directions, it is often appropriate to use two cosine-corrected photocells placed back to back, or to make two separate measurements with the same photocell. This approach is only strictly correct if the target develops predominantly in two dimensions in a plane at right angles to the incident radiation. If the target is predominantly three-dimensional, then a spherical response receiver with a uniform response to all incidence angles is preferable.

One of the main problems with spherical response receivers can be self-shading. It is therefore important that the receiver should be as small as possible relative to the irradiation environment. With some exceptions, many commercially available receivers are too large for many applications. If an appropriate spherical receiver with uniform response cannot be

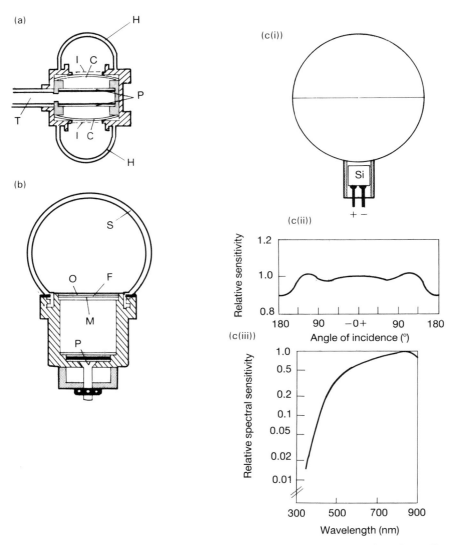

**Fig. 5.4** Examples of spherical response receivers. (a) Receiver with two selenium photocells. H = hemispherical opal glass domes, C = convex opal glasses, I = iris diaphragms, P = photocells, T = tube which serves as handle through which the photocell wires go to the measuring instrument. (Redrawn from Wassink and Vann der Scheer, 1951.) (b) Receiver with single dome and photocell. S = opal glass sphere, O = flat opal glass, F = Schott BG21 (2 mm) filter, M = mosaic spectral sensitivity correction filter, P = photocell below protective clear glass plate. (Redrawn from Šetlík and Kubín, 1966.) (c(i)) Cross section of spherical response receiver constructed from two white Plexiglas hemispheres. (c(ii)) Relative spatial sensitivity. (c(iii)) Relative spectral sensitivity. (Redrawn from Hartmann, 1977.)

purchased there are several basic designs which are not difficult to manufacture. One of the simplest is constructed by cementing the bases of two hemispherical white Perspex or Plexiglas bowls together. The hemispheres can be made from flat sheet, heated in oil until flexible, then formed over a warmed sphere of appropriate diameter. The photocell is either located within a tube which is directly attached to the sphere, or fixed within a hole drilled in the sphere. In both cases, an excellent spherical receiver is obtained, although the sensitivity of the photocell is severely reduced.

There are several items which can be adapted to construct spherical response receivers. The main requirements are that they should be spherical, translucent, preferably diffusing and of appropriate size. One of the most convenient items is a white table tennis ball which has a nominal diameter of 38 mm. The photodetector can be attached directly to the ball, but a preferable approach is to use encased reinforced fibre-optics so that the detector is remote from the receiver. This has the advantages of reducing shading effects and of making the receiver more manoeuvrable in restricted spaces.

The deviation from a perfect response in a spherical response receiver, and to a lesser extent in a cosine response receiver, can be improved by reducing the amount of radiation admitted in the areas of high sensitivity. This can be done by judicious application of a black coating or corrective reflection stripes. The relevant areas are determined during calibration of spatial sensitivity, but are usually opposite the area where the photocell or fibre-optics are attached.


### 5.6.4.  Receiver calibration

The use of a different receiver, or the modification of a standard receiver, can result in three substantial changes in the radiation measurements made by the original instrument. All of these changes must be accounted for during calibration. First, there will be a change in the absolute sensitivity of the instrument. Second, there may be a change in spectral sensitivity. Third, there will be a change in spatial sensitivity. Most manufacturers supply accurate descriptions of these factors, but when purpose-made receivers are used, these potential errors must be measured directly. A standard of spectral irradiance (i.e. energy flux) is required to determine the absolute spectral sensitivity. These procedures are described in Chapter 3 and discussion is limited here to the procedures for determining angular variations in the spatial sensitivity of receivers used in action spectroscopy studies.

A calibrated fixed-position lamp whose spectral quality can be varied is used as a source of radiation. An uncalibrated lamp can be used if only spatial sensitivity measurements are required, but it is important that a stable power supply is used to ensure that the radiation output does not vary during the measurements. The radiation emitted by the lamp must be collimated to restrict the incident angle of the incident radiation at the receiver to less than $10°$, and preferably to around $5°$. The calibrated source can then be used to measure both the spatial and spectral sensitivity of normal incidence, cosine response and spherical response receivers, whereas the uncalibrated source can only be used to measure spatial sensitivity. The receiver is fixed in a holder which allows its rotation in the horizontal and vertical planes; both planes are calibrated in degrees.

The response of a cosine-corrected receiver is measured by first recording the detector output produced by radiation arriving at normal incidence to the receiver. The detector output is then measured as the angle of incidence is moved away from normal incidence in $5°$ stages. The same procedure is repeated after the receiver has been turned about its own axis in 30 or $45°$ steps to ensure that no significant variation exists at other azimuth angles. The receiver's error is then expressed as the percentage deviation in the observed response from that expected with a true cosine response (see Fig. 5.3). The spherical response receiver is calibrated in the same way as the cosine-corrected receiver, except that measurements must be made through $360°$ rather than $180°$.

## 5.7.  Error

The radiation measurements made in action spectroscopy are subject to a wide range of sources of error. The most serious of these can usually be traced back to absolute calibration error and to operator error. Calibration error is usually caused by lack of calibration rather than careless or incorrect calibration. The main reasons for incorrect calibration are usually standard lamp inaccuracy, lamp current inaccuracy, incorrect alignment and excessive stray radiation.

The main sources of operator error are using inappropriate equipment, using uncalibrated equipment, using a different system for calibration from that used for measurement, shading or reflection of radiation by equipment or the operator, incorrect positioning of the receiver, incorrect interconversion of units, dust or moisture modifying the true radiation measurement and inadequate user control of input voltage, zero drift, dark current and temperature. Many of these sources of error can be so serious as to make the radiation measurements meaningless. On the other hand, if the user understands the limitations of the system, operator error can be reduced to negligible proportions.

There are several other sources of error which can be divided into three main areas. These are receiver, detector and indication or readout error. All of these can be minimized by relatively simple precautions. Receiver error has been covered in Section 5.6. Detector error derives mainly from fatigue (slow loss of sensitivity during exposure), non-linearity (inability to follow progressive changes in energy flux), response time (inability to follow rapid changes in energy flux) and temperature. Some detectors exhibit relative error in the spectral response, which is the deviation in the spectral sensitivity of the detector from its designed response. Detectors which are fitted with filters designed to have a tailored spectral response within a specific waveband usually suffer from error in the relative spectral response and are usually unsuitable for action spectroscopy studies. Readout or indication error is the difference between the instrument's readout value and the true value. This problem is usually overcome by using a stable line or battery input voltage, avoiding high humidity and temperature variations, ensuring good electrical connections and by careful control of zero drift and changes in dark current.

The total error in a measurement system is calculated from the various sources of error in the system and is expressed as the square root of the sum of squares of each individual error. This numerical description of radiation measurement error cannot, of course, include a precise definition of operator error. Operator error is best defined by a comprehensive description of the equipment and techniques used.

# References

Dlugos, H. G. (1958). *Lichttechnik* **10**, 565–567.

Eckhardt, G. (1965). *Lichttechnik* **17**, 110A–113A.

Hartmann, K. M. (1977). *In* "Biophysik" (W. Hoppe, W. Lohmann, H. Markl & H. Ziegler, Eds), pp. 197–222. Springer, Berlin.

Kubín, Š. (1971). *In* "Plant Photosynthetic Production. Manual of Methods" (Z. Sestak, J. Catsky & P. G. Jarvis, Eds), pp. 702–765. Junk, The Hague.

Šetlik, I. and Kubín, Š. (1966). *Acta. Univ. Carol., Biol., Suppl.* **1/2**, 77–88.

Wassink, E. C. and Van der Scheer, C. (1951). *Meded. Landbouwhogesch. (Wageningen)* **51**, 175–183.

# 6

# Applications of Lasers in Photobiology and Photochemistry

D. PHILLIPS

*The Royal Institution*
*21 Albemarle Street*
*London W1X 4BS, UK*

## 6.1. Introduction

As in many fields, the introduction of lasers has caused quantum leaps forward in the methodology through which biological systems can be investigated. Questions may now be asked with some prospect of sensible answers which were not feasible before the advent of reliable laser systems. It is the purpose of this article to describe the experimental techniques using principally pulsed laser excitation which either have been used in the study of the excited states of biological molecules or have great potential uses. The special properties of laser radiation which make these devices of such unique usefulness are summarized in Table 6.1, with some commercial and scientific applications.

Table 6.2 shows some commonly available commercial lasers, with their wavelengths. We shall concentrate here on visible, UV and near-IR lasers;

**Table 6.1** Some applications of laser light

| Property | Typical applications |
|---|---|
| Intensity | Tissue destruction, bleaching |
| Coherence | Holography (information storage, non-destructive testing, entertainment), quantum beats |
| Short pulse duration | Communications, study of very fast processes (photochemistry, photo-biology, ballistics) |
| Monochromaticity | Selective excitation, tissue destruction |
| Well-collimated beam | Laser microscopy |
| Plane-polarized beam | Polarization studies, orientation of probe molecules in biology, polymers |

RADIATION MEASUREMENT IN PHOTOBIOLOGY
ISBN 0–12–215840–7

**Table 6.2**   Some common lasers (with wavelengths given)

| Lasing medium | | Type | Wavelength(s) |
|---|---|---|---|
| HF | | Chemical, gas, IR | 2.6–3.0 $\mu$m |
| Diode | | Solid state, IR, low power, CW | 780 nm → 1.55 $\mu$m. Fixed frequency, 2–30 $\mu$m tunable |
| CO | | Gas, IR | 4.7–6.6 $\mu$m |
| $CO_2$ | | Gas, IR, high power | 10.6 $\mu$m |
| Iodine | | Photochemical, gas, IR | 1.315 $\mu$m |
| $Nd^{3+}$ | | Solid state, pulsed, IR harmonics | 1060 nm (530, 265, 353 nm) |
| Ruby | | Solid state, pulsed | 694 nm, (347 nm harmonic) |
| $Kr^+$ | | Gas, CW | 647.1 nm, 530.9 nm |
| He/Ne | | Gas, CW | 632.1 nm, 1150 nm |
| Gold vapour | | Gas, pulsed | 627.8 nm |
| Copper vapour | | Gas, pulsed | 510.6 nm, 578.2 nm |
| $Ar^+$ | | Gas, CW | 514.5 nm, 496.5 nm. 488.0 nm, 476.5 nm, 457.9 nm |
| $N_2$ | | Gas, pulsed | 337.1 nm |
| Excimer | XeF | Gas, UV, pulsed | 351 nm |
| | XeCl | Gas, UV, pulsed | 308 nm |
| | KrF | Gas, UV, pulsed | 248 nm |
| | ArF | Gas, UV, pulsed | 193 nm |
| | $F_2$ | Gas, UV, pulsed | 158 nm |
| Dye | | Flash lamp pumped or pumped by laser | 200–1000 nm with range of dyes (harmonics) |

those which promote *electronic* transitions in molecules. The fates of such excited polyatomic molecules can be discussed with reference to Fig. 6.1, the familiar Jablonski diagram. Following population of the first excited electronic state, $S_1$, the unimolecular electronic relaxation processes competing with fluorescence are intersystem crossing to the triplet manifold (ISC), internal conversion to the ground state (IC) and photochemical reaction. The time domains in which these and some other physical phenomena occur are summarized in Fig. 6.2, together with the limitations of currently available pulsed laser technology. In condensed media vibrational relaxation occurs on a picosecond timescale, and thus only chemical processes with rate constants in excess of $10^{12}$ s$^{-1}$ will compete. Subsequent to excitation therefore vibrational relaxation is usually complete before electronic relaxation. Internal conversion is usually faster at higher excess energies, but is not of great importance for lower lying vibrational levels of

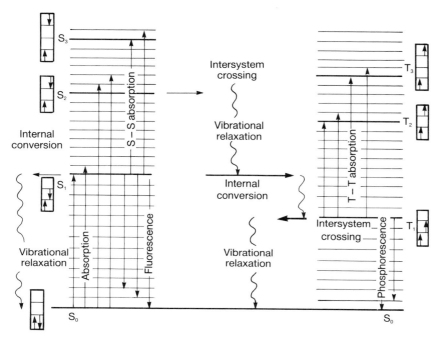

**Fig. 6.1** Jablonski state diagram depicting the fates of photoexcited polyatomic molecules.

the first excited singlet state. The principal process competing with fluorescence is therefore intersystem crossing to the triplet manifold of levels.

The identification of such intermediate states in photochemical reactions including biological systems through spectroscopy and quantitative estimates of reaction rates through the monitoring of the time dependence of the concentration of intermediates are vital pieces of information the gathering of which has been greatly facilitated by the development of pulsed lasers. Reference to Fig. 6.1 indicates by what means the spectroscopy and kinetic studies can be carried out. Clearly, the concentration of excited states must be monitored by some optical technique, and what is available to the experimenter includes the following: (a) emission (fluorescence or phosphorescence); (b) absorption; (c) diffuse reflectance; (d) scattering phenomena, e.g. Raman; (e) transient holographic grating methods. Each of these will be discussed briefly below, with reference to biological systems. Some of the applications of pulsed lasers in biochemistry are shown in Table 6.3.

Monitoring of excited state populations does not merely give information about the intramolecular decay paths of excited states during the course of

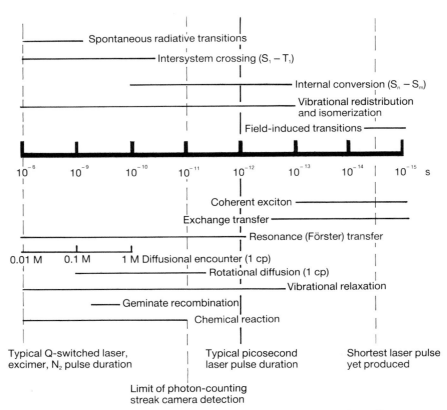

**Fig. 6.2** Some physical and chemical processes which occur on the $10^{-8}$–$10^{-15}$ s timescale. Processes above the horizontal line are intramolecular; those below external to the molecules.

photochemical and photophysical events, since other processes such as diffusion (translational or rotational) may be monitored by using the decay of the excited state molecule as a "clock". In Fig. 6.2, the rates of all of the processes below the centre line may in principle be measured by appropriate choice of probe molecule, the lifetime of which is dictated by the rates of the intramolecular processes shown above the line. Of course, the monitoring of excited state populations requires an excitation source which is short compared with the lifetime of the state of the molecule being monitored. Fig. 6.2 shows that given the availability of lasers producing 10 fs (1 fs = $10^{-15}$ s) pulses, a huge range of timescales is now available. If the excitation of the assembly of absorbing molecules is achieved by plane-polarized radiation, the directional properties of the absorption transition dipole dictate that an instantaneous anisotropy is created in the system

**Table 6.3** Some applications of pulsed lasers in biochemistry

| Topic | Applications | Methods |
|---|---|---|
| Photosynthesis | Energy transfer in photo-synthetic units | Fluorescence decay using solid state lasers streak camera; synchronously pumped dye lasers, up-conversion; single-photon counting |
| | Electron transfer reactions in reaction centres | Transient absorption |
| Cell membranes | Fluidity of membrane | Fluorescence anisotropy, excimer formation, recovery from photobleaching, solvent relaxation |
| | Polarity of site | Time-resolved spectra, single-photon counting |
| Haemoglobin biochemistry | Structure of picosecond iron porphyrin intermediate | Time-resolved resonance, Raman spectroscopy (TR$^3$) |
| Bacteriorhodopsin photocycles | Short-lived intermediates, structures | TR$^3$ |

through photoselection of those molecules with preferred orientation of the transition moment. This anisotropic distribution will relax back to the isotropic distribution by rotational relaxation, and is in principle easily monitored. Some examples of this will be given. Very brief coverage will be given of the use of narrow-bandpass lasers for electronic spectroscopy of complex polyatomic molecules of biological significance in which the spectroscopy is enhanced by cooling of the molecules to very low temperatures in supersonic expansions.

There are many examples of "photochemistry" driven by lasers, which are of biological importance, and the field is too wide to be given comprehensive coverage here. Instead a brief survey of lasers used medically in this way is given, with specific examples of "photodynamic therapy" (PDT) in which a sensitizing dye is used to target the photodestruction of tumorous tissue.

## 6.2.   Time-resolved Studies

The principal technical problem associated with measurement of light intensity on the nanosecond and shorter timescales is that the response time of the monitoring device, photomultiplier or photodiode for example, is long. Various often quite ingenious solutions to the problem have been devised, but in principle all rely on a distance measurement which can be made very accurately, and which knowing the speed of light can be converted easily to a time measurement. Some examples of techniques will now be given.

### 6.2.1. Fluorescence

Early picosecond measurements were made using a *Kerr-cell shutter*, and low repetition rate $Nd^{3+}$ laser. An example is given in Fig. 6.3 (Porter *et al.*, 1974). Doubled or quadrupled light from the $Nd^{3+}$ laser excites the sample; the resultant fluorescence, however, is not seen by the photo-detectors because the polarizers $P_1$ and $P_2$ are crossed. When a delayed fundamental laser pulse (at 1060 nm) arrives at the Kerr cell containing carbon disulphide, the enormous electric field of the light rotates the plane of polarization of light by $90°$, opening the shutter for the duration of the delayed pulse. By making sequential measurements at different delay times, the intensity of fluorescence as a function of time can be recorded.

A variation of this technique is shown in Fig. 6.4 where the principle of "up-conversion" is used to sample fluorescence intensity (Beddard *et al.*,

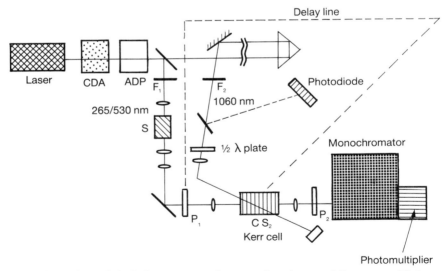

**Fig. 6.3** Early mode-locked arrangement for measuring picosecond fluorescence lifetimes. The laser was a $Nd^{3+}$ in glass laser, with output at 1060 nm, doubled by CDA and quadrupled by ADP crystals to the 530 nm or 265 nm harmonics. The delay line, opening the $CS_2$ Kerr cell shutter, is shown.

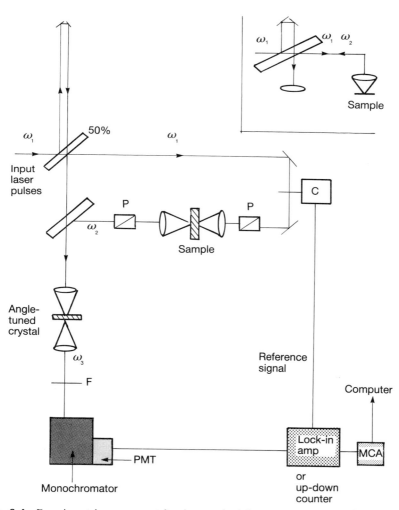

**Fig. 6.4** Experimental arrangement for time-resolved fluorescence up-conversion: F, filters; P, polarizers; C, a sectored disc chopper connected to a lock-in amplifier or photon counter; $\omega_1$, laser beam; $\omega_2$, fluorescence. The crystal is $LiIO_3$ (path length, 1 mm). The sample is contained in a glass cell of path length 1 mm mounted perpendicularly to the exciting beam and spun about an axis parallel to the beam, and the fluorescence is collected from the front face of the cell along the exciting beam axis. In an alternative method (inset) the sample is flowed through a cell or pumped through a nozzle to form a jet, and fluorescence is collected at 180° to the exciting beam.

1981; Doust *et al.*, 1984a). Here a delayed laser pulse is mixed with fluorescence in a crystal which gives an output at the sum of the laser and fluorescence frequencies. Thus fluorescence light at say 900 nm ($\lambda_1$) mixed with 600 nm ($\lambda_2$) laser light gives an output at $\lambda$, given by Equation (1) which in this case is 360 nm, and easily detectable:

$$\frac{1}{\lambda} = \frac{1}{\lambda_1} + \frac{1}{\lambda_2}. \tag{1}$$

Another means of detecting fluorescence photons is the use of a "streak" camera. The principle is very simple: fluorescence photons arriving at the cathode of the tube eject electrons which are accelerated towards a phosphor screen. However, en route they encounter a rapidly changing electric field, which means that electrons passing through this field at different times are deviated to different extents, thus causing a "streak" on the phosphorescent screen. This streak can be calibrated in terms of the arrival time of the photons causing ejection of electrons, and is capable of about 10 ps time resolution.

Two versions of the streak camera system are available: the first operates the camera in the "single shot" mode, in which a single event resulting from a single high power laser pulse can be recorded. The second operates in what is known as synchroscan mode, in which the voltage applied to the deflector plates is linearly ramped at a frequency proportional to the pulse repetition rate. In this mode, high repetition rate lasers can be used, permitting averaging of the recorded signals. A typical layout of a synchroscan system is shown in Fig. 6.5. It is based on a cavity-dumped, synchronously pumped dye laser. The optical delay enables recording of the excitation pulse and fluorescence emission to be recorded in the same time frame, for ease of data analysis. A photodiode detector enables synchronization of the laser pulses with the scanning of the streak camera deflector plates. The "streaked" emission from the phosphor screen is detected using an optical multichannel analyser (OMA), utilizing a Vidicon, Reticon or CCD camera multichannel detector.

An example will now be cited of the use of a streak camera system to study events in *photosynthesis*. The overall pattern of natural photosynthesis (Doust *et al.*, 1984b) consists of a chain of electron transfer reactions which, in green plants and algae, result at the electron donor end in the oxidation of water to oxygen, and at the electron acceptor end in reduction processes such as the conversion of carbon dioxide to carbohydrate, of nitrogen to ammonia or of protons to hydrogen. These involve two photochemical steps and two separate, but linked, photochemical systems. Photosystem II oxidizes water and produces a reduced intermediate (e.g. a hydroquinone). Photosystem I oxidizes the hydroquinone and produces the reducing equivalents to make carbohydrate, etc. Each photosystem requires at least one photon for each electron transfer so that two photons are utilized for one complete electron transfer from the oxidizing to the reducing side. Each of the two photosystems consists of two important parts. First, there is an array of pigment molecules, in the light-harvesting antenna, whose purpose is to collect the light energy and transfer it to the reaction centre. The reaction centre consists of a special pigment molecule

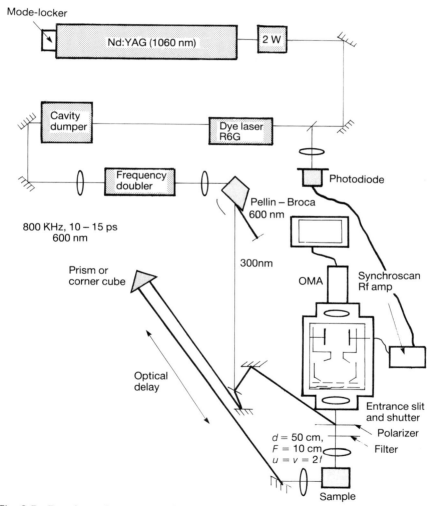

**Fig. 6.5** Experimental arrangement for the study of fluorescence decay using a streak camera detection (see text).

and several redox systems coupled in such a way that the excited pigment receptor can bring about electron transfers sequentially along the redox chain.

The studies quoted here refer to the light harvesting antenna in photosystem II in an alga, *Porphyridium cruentum*, in which the arrangement of three of the four protein pigment complexes is like the shells of an onion, with the shortest wavelength absorber, phycoerythrin, on the outside, then phycocyanin, then allophycocyanin (Porter *et al.*, 1978). The last pigment to which the energy is finally transferred is chlorophyll which is in the membrane to which the phycobilisome is attached. In the experiment, the 530 nm pulse from a frequency-doubled Nd laser was used to excite the phycoerythrin, being relatively little absorbed by the other pigments. All four pigment fluorescences can be fairly well separated by interference filters transmitting 10 nm wide bands centred at 576, 640, 661 and 685 nm. The results of these experiments are summarized in Table 6.4. The sequential rise and decay of the fluorescence at the four wavelengths is clearly shown and the times were independent of pulse intensity over a tenfold intensity range. When the phycobilisomes were removed from the membrane, the lifetime of the allophycocyanin was increased from 118 ps to 4 ns since the chlorophyll acceptor was no longer present, but the kinetics of the fluorescence of the other two pigments were unaffected, showing that the energy transfer was sequential and direct transfer from the outer pigments to chlorophyll was not important.

The decay of phycoerythrin fluorescence was not exponential but fitted well the decay law $I(t) = I_0 \exp(-2At^{1/2})$. This relation is frequently found to give closer agreement to experimental decays than the exponential law in very fast kinetic studies and as predicted theoretically for the early stages of time-dependent diffusion processes and for Forster-type energy transfer to a random array. Similar measurements have been made using very low excitation intensities of $10^8$ photons $cm^{-2}$ per pulse from a dye laser with time-correlated single-photon counting giving time resolution of about 25 ps (see below) confirming the conclusions above, including the $\exp(-2At^{1/2})$ decay law (Yamazaki *et al.*, 1984).

**Table 6.4** Fluorescence of pigments in *Porphyridium cruentum* alga and in phycobilisomes detached from the membrane (Porter *et al.*, 1978)

|  | Wavelength (nm) | $\tau_{1/2}$(rise)ps Alga | $\tau_{1/2}$(decay) Alga | ps Phycobilisome |
|---|---|---|---|---|
| Phycoerythrin | 578 | 0 | $70 \pm 5$ | $70 \pm 5$ |
| Phycocyanin | 640 | 12 | $90 \pm 10$ | — |
| Allophycocyanin | 660 | 22 | $118 \pm 8$ | 4000 |
| Chlorophyll | 685 | 52 | $175 \pm 10$ | — |

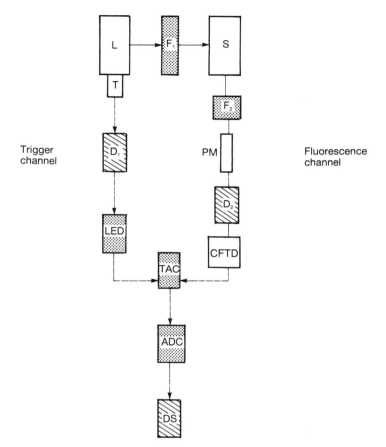

**Fig. 6.6** Block diagram of conventional single-photon counting apparatus. ——— Optical signal; – – – electronic signal. L, excitation source; T, trigger (antenna, fibre-optic and photomultiplier tube, etc.); $F_1$, $F_2$, wavelength selection for excitation and emission; $D_1$, $D_2$, delay lines; LED, leading edge discriminator; PM, photomultiplier; CFTD, constant fraction timing discriminator; TAC, time-to-amplitude converter; ADC, analogue-to-digital converter; DS, data store (multichannel analyser or computer).

The basic principles of time-correlated single-photon counting have been the subjects of many reviews (Ware, 1971; Yguerabide, 1972; Badea and Brand, 1979; Knight and Selinger, 1973; Isenberg, 1984; Pultney, 1972; O'Connor and Philips, 1984; Phillips *et al.*, 1985). The method relies on the basic concept that the probability distribution for emission of a single photon following excitation gives the actual intensity against time distribution of all photons emitted, thus by sampling the time of single photon emission following a large number of excitation pulses, the probability distribution is created.

The experiment is carried out as follows, with reference to Fig. 6.6. A trigger T, which could be a photomultiplier, an antenna pick-up or a logical synchronizing pulse from the electronics pulsing the excitation source, generates an electrical pulse at a time exactly correlated with the time of generation of the optical pulse. The trigger pulse is routed, usually through a discriminator, to start input of the time-to-amplitude converter (TAC) which initiates a linear voltage ramp. In the meantime the optical pulse excites the sample which subsequently fluoresces. An aperture is adjusted so that at most one photon is "detected" for each exciting event. The signal resulting from this photon stops the voltage ramp in the TAC which puts out a pulse, the amplitude of which is proportional to the final ramp voltage, and hence to the time difference between START and STOP pulses. The TAC output pulse is given a numerical value in the analogue-to-digital converter and a count is stored in the data storage device in an address corresponding to that number. Excitation and data storage are repeated in this way until the histogram of number of counts against address number in the storage device has enough data so that it represents, to some required precision, the fluorescence decay curve of the sample. If deconvolution is necessary, the time profile of the excitation pulse is collected in the same way by replacing the sample by a light scatterer. The technique is capable of prodigious sensitivity and extremely high signal-to-noise ratio. With the advent of microchannel plate detectors, the instrument response function can be as low as 80 ps, giving an ultimate time resolution of 10 ps.

The requirements in lasers for TCSPC applications are high repetition rate tunable pulses of short duration. Mode-locked continuous wave (CW) lasers fall into this category providing pulse duration of < 10 ps (dye lasers), 200 ps (ion lasers) or 100 ps (Nd:YAG lasers). When used in the mode-locked mode, the repetition rate is fixed by laser cavity length and is normally of the order of 76 or 82 MHz. Such high repetition rates are too fast even for TCSPC and therefore a reduction in rate is required. Extracavity devices, such as a *Pockels' cell*, have been used, but suffer from an inherent large RF interference signal associated with the high voltage changes, a repetition rate which is too low and inefficient use of the available pulses. The most commonly used device for controlling the pulse repetition rate is the acoustic-optic, intracavity Bragg cell, known as a *cavity dumper*. Such a device has been used with mode-locked argon and krypton ion lasers and synchronously pumped, mode-locked dye lasers. Depending upon the electronic drive being used, repetition rates are continuously selectable from, for example, 76 MHz to single shot. When used in unmode-locked CW laser systems, the cavity dumper can provide the same flexibility and stability in pulse output, providing pulse durations of ~ 10 ns, which have proved useful in measuring lifetimes down to 1 ns. Until recently, the most commonly used system was the cavity-dumped, mode-locked dye laser synchronously pumped by a mode-locked argon ion laser. Whilst such a system provides the required repetition rate and pulse duration, the wavelength tunability was limited to dyes that could be pumped by the 514.5 nm argon ion lasing line, 560–760 nm, or 280–380 nm if the output was frequency doubled. Mode-locking of the 364 nm UV line of an argon laser has recently permitted access to dyes which "lase" in the 380–500 nm region of the spectrum. However, the most practical system currently available uses a mode-locked CW Nd:YAG laser as a pump source. By frequency doubling the 1064 nm output to 532 nm or frequency tripling to 355 nm, a large range of dyes from the UV to the IR are accessible. A system currently in use in our laboratory is shown schematically in Fig. 6.7. The intensity of such a low-powered laser source (e.g. 100 mW at 600 nm) nevertheless represents $6 \times 10^{16}$ photons per second, greatly in excess of conventional light sources.

The changes in dispersed fluorescence as a function of time can provide

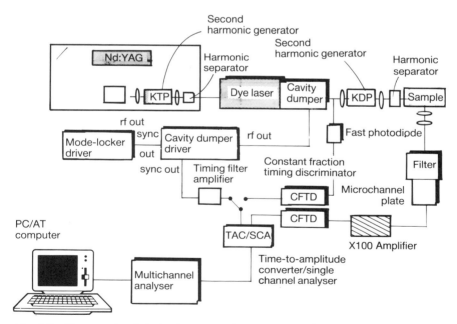

**Fig. 6.7**  Time-correlated single-photon counting spectrometer based on a CW mode-locked Nd:YAG laser.

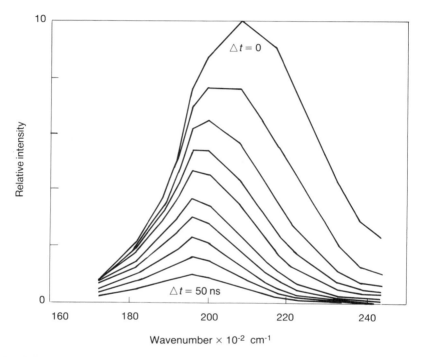

**Fig. 6.8** Time-resolved emission spectra (TRES) of dansyl propanalol in lipid vesicle (see text). As time progresses on a nanosecond timescale, note progressive red shift in spectral profile and collapse in intensity, typical of solvent relaxation occurring on same timescale as the electronic relaxation of the fluorophore. Spectra are identified for a time delay of 0 ns and 50 ns. The other curves are at regular time intervals between these limits.

invaluable information in complex systems, including the identification of a number of emitting species, and their temporal or kinetic relationship. A rapid method of providing uncorrected, that is "convolved", spectra exists with the time-correlated single-photon counting method, using upper and lower voltage discriminators on the time-to-amplitude converter to provide a "time" gate. An example of the use of this technique to yield data on biological systems is given in Fig. 6.8, in which the change in spectral profile

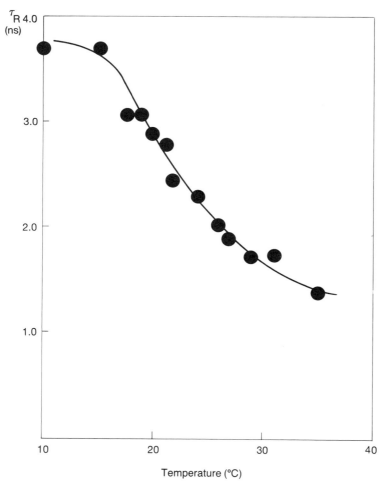

**Fig. 6.9**  Effect of temperature on the rate of the solvent relaxation $\tau_R$ of dansyl propanalol in dimyristoyl phosphatidylcholine vesicles. The phase transition of the lipid is clearly revealed by this technique.

as a function of time of a fluorescent probe based upon the "dansyl" moiety is shown in a biomembrane (Stubbs *et al.*, 1985). The reduction in intensity and progressive red shift of the fluorescence spectrum as a function of time are the consequences of electronic relaxation of the probe occurring at the same time as solvent reorganization around the changed excited state dipole. By monitoring such solvent relaxation by laser-based methods an

alternative method is made available for the probing of membrane micro-
viscosity. A typical result is given in Fig. 6.9, which clearly shows the phase
transition in dimyristoyl phosphatidylcholine vesicles probed by dansyl
propanalol (Stubbs *et al.*, 1985).

More conventional measurements of motion of molecules rely upon the
monitoring of fluorescence polarized parallel $I_\parallel$ and perpendicular $I_\perp$ to the
plane of polarization of exciting laser radiation, and deriving the anisotropy
$r(t)$ from these measurements (Equation (2)):

$$r(t) = [I_\parallel(t) - I_\perp(t)]/[I_\parallel(t) + 2I_\perp(t)] = D(t)/S(t). \tag{2}$$

The different approaches to analysing the time dependence of the anisotropy generally arise from different methods by which deconvolutions of
$I_\parallel(t)$ and $I_\perp(t)$ are translated into a deconvoluted $r(t)$. For example, the
rotational parameters can be extracted by (i) individually deconvolving
$I_\parallel(t)$, (ii) individually deconvolving $I_\perp(t)$, (iii) deconvolving $D(t)$, (iv)
deconvolving both $D(t)$ and $S(t)$ and then reconstructing $r(t)$, (v) simultaneously fitting $I_\parallel(t)$ and $I_\perp(t)$, (vi) simultaneously analysing several decay

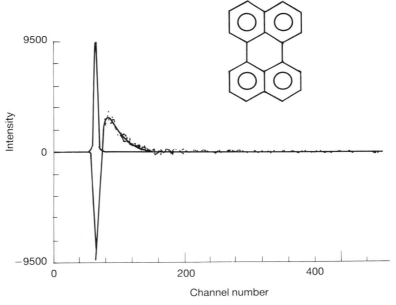

**Fig. 6.10** Polarization of fluorescence, as a function of time, of perylene (shown on
diagram) in glycerol following excitation by second harmonic argon ion laser at 257.25 nm.
Anisotropy, $r$, is defined in terms of intensities of fluorescence parallel ($I_\parallel$) and perpendicular
($I_\perp$) to plane of polarization of laser exciting light, as $r = (I_\parallel - I_\perp)/(I_\parallel - 2I_\perp)$. Ordinate is $r(t)$
channel number, linear in time.

curves ("global analysis") (Knutson *et al.*, 1983), etc. These and other methods have been discussed in some detail by Cross and Fleming (1984) and Christensen *et al.* (1986). At this point, there is no general agreement on which of these methods is most accurate, most efficient, least subject to typical systematic errors, etc. As an example of the detailed information about rotational motion which is accessible by such experiments in favourable cases, Fig. 6.10 shows the time dependence of anisotropy for the probe molecule perylene in the viscous medium glycerol and water. The experimental curve can be mimicked by assuming two independent rotations of the probe with different diffusion coefficients.

The advances in time-resolved techniques alluded to above have fostered a re-examination of theories of the rotational motions of molecules in liquids. Models considered include: the anisotropic motion of unsymmetrical fluorophores (Tao, 1969; Belford *et al.*, 1972); the internal motions of probes relative to the overall movement with respect to their surroundings; the restricted motion of molecules within membranes (e.g. wobbling, within a cone) (Kinoshita *et al.*, 1977; Szabo, 1984). Analyses of these models point to experimental situations in which the anisotropy can show both multi-exponential and non-exponential decay. Current experimental techniques *are* capable of distinguishing between these different models, but it should be emphasized that to extract accurately a single "average" rotational correlation time demands the same precision of data and analysis as fluorescence decay experiments which exhibit dual exponential decays. Multiple or non-exponential anisotropy experiments are thus near the limits of present capabilities and generally demand favourable combinations of fluorescence and rotational diffusion times.

Although pulse-counting methods have tended to play a dominant role in the investigation of the time-resolved luminescence of synthetic polymers, the complementary technique, that of frequency domain phase-modulation fluorimetry, widely used in biochemistry (Lakowicz *et al.*, 1984a,b), deserves some coverage here. Instead of employing pulsed excitation in this technique the excitation beam is deeply modulated sinusoidally at a frequency comparable to the decay of the sample. Information concerning the decay law of the sample is obtained from the phase shift ($\phi$) and the depth of modulation ($m$) of the emission, both measured relative to the phase and modulation of the incident light. For pulse fluorimetry it is ideal to have a narrow excitation pulse, whereas for phase-modulation fluorimetry, it is ideal to have a wide range of modulation frequencies. Developments in laser technology have made progressively shorter pulses of light available, and this has stimulated growth in the field of pulse fluorimetry. By contrast, most commercial phase-modulation fluorimeters operate at only two or three modulation frequencies, which limits the

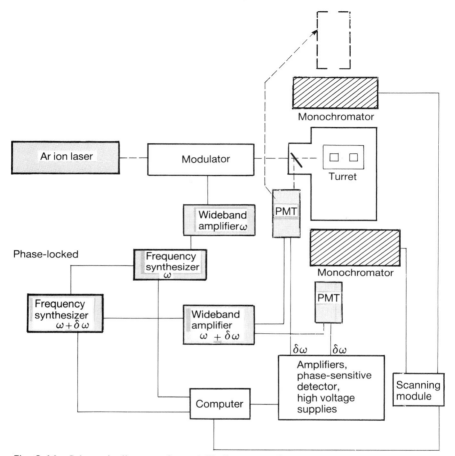

**Fig. 6.11** Schematic diagram of a variable-frequency phase modulation fluorimeter, based upon laser excitation (see text).

information content and resolving power. However with the development of new modulators, phase-modulation methods using lasers have now come into their own (Gratton *et al.*, 1984; Lakowicz *et al.*, 1984a,b, 1986; Lakowicz and Cherek, 1985; Lakowicz and Maliwal, 1985). An apparatus using this method is shown in Fig. 6.11. In this apparatus the output of a continuous laser is intensity modulated in an approximately sinusoidal manner using an electro-optic modulator which is placed between crossed polarizers. The modulated intensity is used to excite the sample, and the emission is observed using a photomultiplier. One photomultiplier is used as a phase reference to monitor the phase of the incident light. Detection of the emission is performed using the cross-correlation method. Specifically, the gain of the PMTs are modulated at the frequency $f + 25$ Hz, where $f$ is the modulation frequency of the incident light. The phase and modulation of the 25 Hz cross-correlation frequency is measured by a time-interval counter and a ratio digital voltmeter. A variety of CW lasers, including argon ion and helium−cadmium, may be used for excitation.

### 6.2.2.   Pump and probe experiments

Monitoring of the electronic (and in some cases vibrational) absorption spectrum of transient species following pulsed excitation permits both the absorption spectrum and kinetics of formation and decay of the transient to be measured. The technique, flash photolysis, has been in existence more than 40 years; the use of lasers as excitation source is already 25 years old. The field has been reviewed very thoroughly recently (West, 1986). Nanosecond measurements using doubled, tripled and quadrupled $Nd^{3+}$ lasers, ruby lasers and more recently excimer lasers are now routine. Here we highlight picosecond measurements, for which experimental techniques are much more exacting.

A laser system which allows picosecond transient absorption measurements to be made is shown in Fig. 6.12 (Gore *et al.*, 1986). Light from a synchronously pumped dye laser in the form of picosecond pulses is amplified in four stages, A1−A4, with an Nd/YAG laser pumping the amplifying dyes. This yields pulses of around 1 ps duration, probe energy $>1$ mJ, wavelength 595−605 nm, at a pulse repetition rate of 10 Hz. Any transient species produced in the sample chamber (SC) are interrogated with a delayed pulse of white light, which is produced by continuum generation, i.e. white light produced by passage of an intense laser pulse through a liquid, often water. The time delay between pump and white light probe can be any time between 0.1 ps and 12 ns. Light absorbed from the white light continuum is measured by comparison of beams passed through sample cell and reference cells, analysed with a spectrograph or monochromator, and detected using a vidicon, a detector array in the focal plane of the monochromator which allows simultaneous recording of intensities at a whole range of wavelengths. Figure 6.13 shows results of some such pump and probe experiments on the photosynthetic units from pea chloroplasts. Two main spectral features can be observed in Fig. 6.13, one at 690 nm and one at 700 nm. That at 690 nm is dominant in early times, decays with a lifetime of ca 15−20 ps and undergoes a progressive blue shift, finally being centred at 675 nm. The second feature, centred at 700 nm, is very much narrower and occurs only in the spectra of samples in which

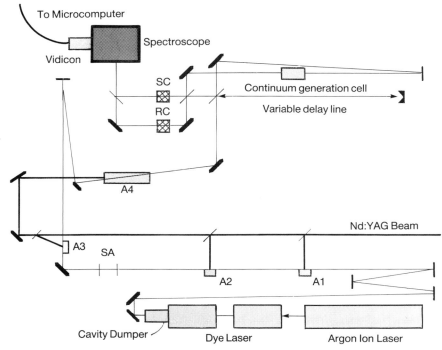

**Fig. 6.12** Picosecond transient absorption spectrometer, utilizing the four-stage amplification of a synchronously pumped dye laser. SA, saturable absorber; A1, amplifier cell No. 1;
etc. SC, sample cell; RC, reference cell.

chlorophyll P700 is chemically reduced. The signal at 690 nm can be
attributed to the excitation of antenna chlorophyll to the singlet state while
the signal at 700 nm can be attributed to the photo-oxidation of P700
molecules. Such experiments provide an important means of studying the
very fast chemical processes which occur very early in photosynthesis.

Transient absorption measurements require that the sample under
investigation be transparent, and non-scattering. For many applications,
particularly those in the field of polymer science, this condition may not be
met. A technique has been developed which permits recording of the diffuse
reflectance spectra of transients produced upon laser flash photolysis
(Wilkinson *et al.*, 1986).

Another detection technique, which is a variation on a transient absorption method, utilizes a transient holographic grating to study the time
dependence of the polarization of absorbing molecules (Moog *et al.*, 1982;
Hyde *et al.*, 1986). These time-resolved optical experiments rely on a short

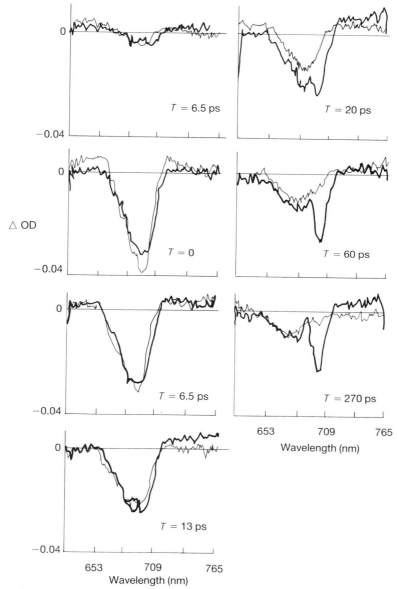

**Fig. 6.13** Transient absorption spectra of photosystem I reaction centres for pea chloroplasts. (━━━) Spectra with chlorophyll P700 chemically reduced; (────) with P700 chemically oxidized. The delay times in picoseconds are shown.

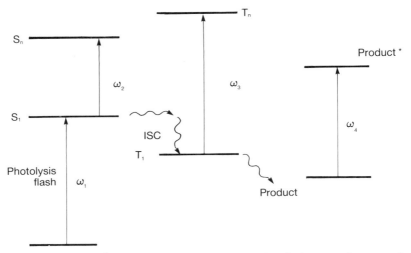

**Fig. 6.14** Basis of TR$^3$ measurement. Pump laser $\omega_1$ excites singlet state of compound. As probe laser scans $t$ through different resonances $\omega_2$, $\omega_3$, $\omega_4$, the non-linear resonant enhancement of Raman scattering from, in this case, the $S_1$ state, the $T_1$ state and a product can be selectively enhanced. Spectra derived are much more clearly separated than absorption spectra.

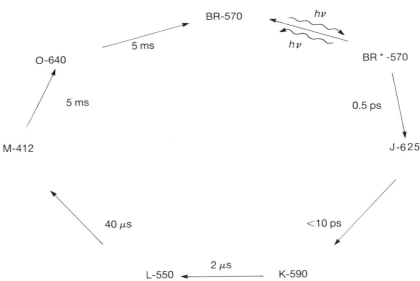

**Fig. 6.15** The bacteriorhodopsin photocycle. Each intermediate is identified by a letter followed by a number which corresponds to the $\lambda_{max}$ in nm of its visible absorption spectrum. Only the initial change BR-570 → BR$^*$-570 is light driven, the remainder of the cycle being thermal, with time intervals shown.

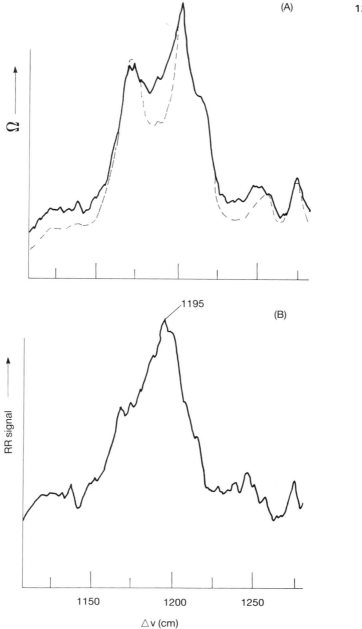

**Fig. 6.16** TR³ spectra of bacteriorhodopsin at 40 ps delay. Panel A contains the TR³ spectrum with a 590 nm probe pulse (4 mW, 8 ps, 1 MHz) 40 ps after excitation at 565 nm (20 mW, 8 ps, 1 MHz). The dashed trace is the BR-570 spectrum obtained with the probe pulse only. Panel B shows the resonance Raman spectrum of the K-590 intermediate obtained by subtraction of the BR-570 contribution from the TR³ spectrum.

pulse of exciting plane-polarized light from a laser to photoselect chromophores which have their transition dipoles orientated in the same direction as the polarization of the exciting radiation, giving a non-random orientational distribution of excited state dipoles, which in the case of polymers and biopolymers will randomize in time due to motions of the polymer chains to which the chromophores are attached.

In the transient grating experiment, optical interference between two crossed laser pulses creates a spatially periodic intensity pattern in an absorbing sample. This results in a spatial grating of excited states which then diffracts a third (probe) beam brought into the sample at some later time. The two observable experimental quantities are the intensity of the diffracted signal for the probe beam polarized parallel ($T_{\parallel}(t)$) and perpendicular ($T_{\perp}(t)$) to the excitation radiation (cf fluorescence polarization above).

Transient absorption (and reflectance) spectra of complex polyatomic molecules are in principle broad and featureless, and do not yield structural formation. Vibrational spectra are capable of yielding structural details and recently, time-resolved resonance Raman ($TR^3$) spectroscopy has been used to advantage to study transients in photochemical systems (Phillips *et al.*, 1986). The basis of the method is summarized in Fig. 6.14. Nanosecond measurements are now routine, but extension of such measurements into the picosecond time domain has been facilitated by the advent of high-repetition-rate copper vapour lasers, for example. As an example of the use of this technique Fig. 6.15 shows schematically the bacteriorhodopsin photocycle. In Fig. 6.16 the transient Raman spectrum of species K-590 is shown (Atkinson *et al.*, in press). The technique offers much promise in studies of the structures of intermediates in photochemical and photobiological systems.

## 6.3.  Spectroscopic Studies

It would be inappropriate within the confines of this brief review to attempt to cover the vast field of laser spectroscopy, even if attention were to be confined to photochemical and photobiological systems of particular interest. Instead, attention will be drawn very briefly to a technique which, albeit improbable, offers hope in the interpretation and simplification of electronic spectroscopy of complex polyatomic molecules of biological interest. Under the envelope of the electronic absorption band of a complex polyatomic molecule, such as that shown in Fig. 6.17 (Phillips, 1988), there are a myriad of overlapping individual narrow resonances, irresolvable even by narrow-bandpass laser excitation. The situation can be greatly improved

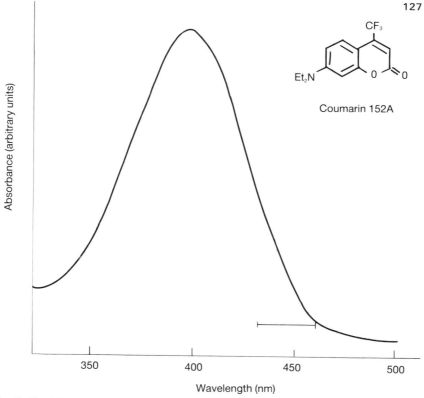

**Fig. 6.17** Absorption spectrum of dyestuff coumarin 152A (structure shown) in solution phase at room temperature.

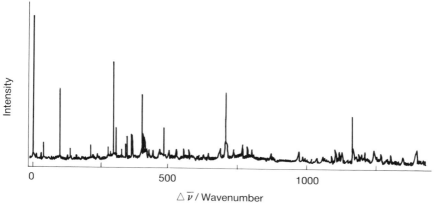

**Fig. 6.18** Supersonic jet fluorescence excitation spectrum, equivalent to an absorption spectrum of the coumarin dye (coumarin 142) for comparison with the solution phase spectrum in Fig. 6.17. Note the sharp resonances which are easily assigned to particular vibrations.

by casting the molecules in the gas phase in isolation to temperatures approaching absolute zero (Levy, 1984; Phillips, 1988). The is achieved by supersonic jet expansion and the result, demonstrated using the same molecule as in Fig. 6.17, is shown in Fig. 6.18 to be dramatic. Such studies have been carried out on indoles (models for tryptophan) (Hager and Wallace, 1983; Phillips and Levy, 1986; Bickel *et al.*, 1987), and tryptophan itself (Bersohn *et al.*, 1984; Rizzo *et al.*, 1986a,b) with great effect. The usual monitoring technique is fluorescence (Rizzo *et al.*, 1986a) excitation, but multiphoton ionization provides a sensitive alternative for non- or weakly fluorescent materials (Rizzo *et al.*, 1986b).

## 6.4. Lasers and Living Biological Systems

The obvious application of lasers in living systems is of course in medicine, the science of which, photomedicine, has been well reviewed recently (Regan and Parrish, 1982). Cryptically, the effects of laser light upon living tissue are summarized in Table 6.5 (Phillips, 1984). The effect of low level light, not necessarily from lasers, is biostimulation (Karu, 1987). This area is controversial, but appears now to be substantiated and may have immense applications. At the highest intensities, vaporization by infra-red lasers is used as a surgical tool, but from a photochemical point of view vaporization by ultraviolet lasers, photoablation, without the dumping of heat into the system promises much, and is being explored actively as a means of correcting eye defects (radial keratectomy), as a means of painless dentistry and for microsurgery and angioplasty.

We concentrate here on a technique of selectively destroying tumour tissue using red lasers, fibre-optics and a sensitizing dye which has some selectivity for tumour tissue. This is a burgeoning field, but as yet there are many unanswered questions, particularly with regard to the mechanism of transport of the dye and selective uptake or retention. Photochemically, the

**Table 6.5**  Interaction of laser light with living tissue

| Energy density $(J.cm^{-2})$ | Biological effect |
|---|---|
| < 4 | Biostimulation |
| > 4 | Biosuppression |
| 40 | Non-thermal cytotoxic phototherapy with sensitizing agents |
| 400 | Photocoagulation |
| 4000 | Vaporization |

AlSPc                                    Haematororphyrin

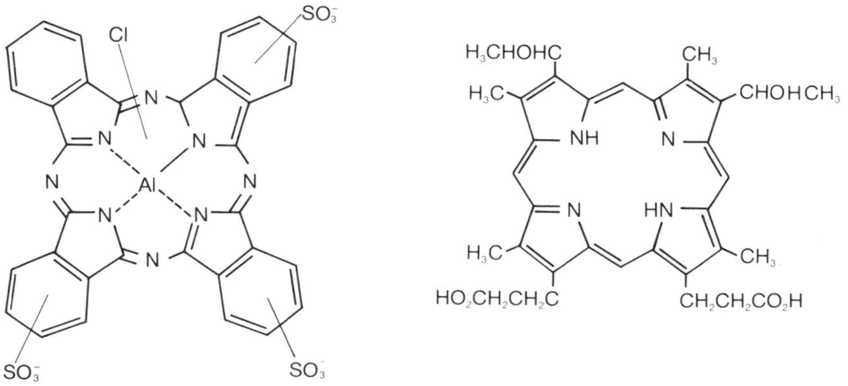

Phthalocyanine

Scheme 6.1.

mode of action of haematoporphyrin derivative (Kessel, 1987) and possibly phthalocyanines (Tralau *et al.*, 1987) (Scheme 6.1), the two main sensitizers used, appears to be through the excitation of singlet oxygen through reactions (3)–(6), although there are other possibilities.

$$S + h\nu \rightarrow {}^1S^* \tag{3}$$

$${}^1S^* \rightarrow S + h\nu \tag{4}$$

$${}^1S^* \rightarrow {}^3S^* \tag{5}$$

$${}^3S^* + O_2 \rightarrow S + {}^1O_2{}^*. \tag{6}$$

A wide variety of tumours have been treated on an experimental basis using the method, which is an extremely vigorous field of research.

The lasers currently used in medical applications are summarized in Table 6.6 (Phillips, 1984).

## 6.5. Conclusions

It will be seen that there are a very large number of applications of lasers in biological studies, only some of which have been touched on in this brief survey which has concentrated upon time-resolved methods, reflecting just one of the unique features of laser radiation. A fuller description of the use of lasers in biochemistry and biology will be found in the recent book by Letokhov (Letokhov, 1987).

**Table 6.6**   Lasers currently used in medicine

| Type | Wavelength | Uses |
|------|-----------|------|
| Argon ion | 514.5 nm | Ophthalmology |
|  | 488 nm | Dermatology |
| Carbon dioxide | ca 10.6 $\mu$m | Gynaecology |
|  |  | Otolaryngology |
|  |  | Neurosurgery |
|  |  | Plastic and general surgery |
| Neodymium:YAG | 1060 nm | Endobronchial surgery |
|  | (530 nm doubled) | Gastroenterology |
| Ruby | 694 nm | Dermatology |
|  | (347 nm doubled) |  |
| Helium–neon | 632.8 nm | Diagnostic |
|  |  | Acupuncture |
| Dye | Tunable, | Photodynamic |
|  | but usually | therapy (PDT) |
|  | 550–800 nm |  |
| Excimer | 158 nm ($F_2$) | Used experimentally, |
|  | 193 nm (ArF) | possible uses in eye surgery, |
|  | 248 nm (KrF) | dentistry, microsurgery. |
|  | 308 nm (XeCl) | Vaporization occurs without |
|  | 351 nm (XeF) | tissue being heated |

## Acknowledgements

Gratitude is expressed to SERC, Imperial Cancer Research Fund and US Army European Research Office for generous support of work included in this report.

## References

Atkinson, G. H., Brack, T. L., Blanchard, D. & Rumbles, G. Picosecond time-resolved raman spectroscopy of the initial *trans* to *cis* isomerization in the Bacteriorhodopsin Photocycle, *Chem. Phys.* (in press) communicated privately by G. Rumbles.

Badea, M. G. & Brand, L. (1979). Time Resolved Fluorescence Measurements. *Meth. Enzymol.* **61**, 378.

Beddard, G. S., Doust, T. A. M., Meech, S. R. & Phillips, D. (1981). Synchronously Pumped Dye Laser in Fluorescence Decay Measurements of Molecular Motion, *J. Photochem.* **17**, 427.

Belford, G. G., Belford, R. L. & Weber, G. (1972). 'Dynamics of Fluorescence Polarisation in Macromolecules', *Proc. Natl. Acad. Sci. USA* **69**, 1392.

Bersohn, R., Even, V. & Jortner, J. (1984). 'Fluorescence excitation spectra of indole, 3-methyl indole and 3-indole acetic acid in supersonic jets', *J. Chem. Phys.* **80**, 1050.

Bickel, G. A., Leach, G. W., Demmer, D. R., Hager, J. W. & Wallace, S. C. (1987). The torsional spectra of jet cooled methyl substituted indoles in the ground and first excited states'. *J. Chem. Phys.* **88**, 1.

Christensen, R. L., Drake, R. C. & Phillips, D. (1986). Time-Resolved Fluorescence Anisotropy of Perylene, *J. Phys. Chem.* **90**, 5960.

Cross, A. J. & Fleming, G. R. (1984). Analysis of Time Resolved Fluorescence Anisotropy Decays, *Biophys. J.* **46**, 45.

Doust, T. A. M., Porter, G. & Phillips, D. (1984a). Picosecond Spectroscopy: Applications in Biochemistry Part I: Techniques. *Biochem. Soc. Trans.* **12**, 630.

Doust, T. A. M., Porter, G. & Phillips, D. (1984b). Picosecond Spectroscopy: Applications in Biochemistry Part II: Applications. *Biochem. Soc. Trans.* **12**, 633.

Gore, B. L., Doust, T. A. M., Giorgi, L. B., Klug, D. R., Ide, J. P., Crystall, B. & Porter, G. (1986). The Design of a Picosecond Flash Spectroscope and its Application to Photosynthesis. *J. Chem. Soc. Faraday II* **82**, 2111.

Gratton, E., Limkeman, M., Lakowicz, J. R., Maliwal, B. P., Cherek, H. & Laczko, G. (1984). 'Resolution of Mixtures of Fluorophores using Variable Frequency Phase and Modulation Data'. *Biophys. J.* **46**, 479.

Hager, J. & Wallace, S. C. (1983). 'Laser spectroscopy and photodynamics of indole and indole van der Waals Molecules in a supersonic beam'. *J. Phys. Chem.* **87**, 2121.

Hyde, P. D., Waldow, D. A., Ediger, M. D., Kitano, T. & Ito, K. (1986). Local Segmental Dynamics of Polyisoprene in Dilute Solution: Picosecond Holographic Grating Experiments, *Macromolecules* **19**, 2533.

Isenberg, I. (1984). Time-Decay Fluorimetry by Counting. In "Biochemical Fluorescence Concepts" (R. F. Chen & H. Edelhoch, Eds), Vol. 1, p. 43. Dekker, New York.

Karu, T. I. (1987). 'Photobiological fundamentals of low-power laser therapy' *IEE J. Quantum Electronics* **QE23**, 1703.

Kessel, D. (1987). 'Tumour Localization and Photosensitization by derivaties of haematoporphyrin: A review' *IEE J. Quantum Electronics* **QE23**, 1718.

Kinoshita, K., Kawato, S. & Ikegami, A. (1977). 'The Theory of Fluorescence Polisation in Membranes'. *Biophys. J.* **20**, 289.

Knight, A. E. W. & Selinger, B. K. (1973). Single Photon Decay Spectroscopy. *Aust. J. Chem.* **26**, 1.

Knutson, J. R., Beecham, J. M. & Brand, L. (1983). Simultaneous Analysis of Multiple Fluorescence Decay Curves: A Global approach. *Chem. Phys. Lett.* **102**, 501.

Lakowicz, J. R. & Cherek, H. (1985). 'Resolution of an Excited State Reaction using Frequency Domain Fluorimetry'. *Chem. Phys. Lett.* **122**, 380.

Lakowicz, J. R. & Maliwal, B. P. (1985). Construction and Performance of a Variable-Frequency Phase-Modulation Fluorimeter. *Biophys. Chem.* **21**, 61.

Lakowicz, J. R., Laczko, G., Cherek, H., Gratton, E. & Limkeman, M. (1984a). 'Analysis of Fluorescence Decay Kinetics from Variable Frequency Phase Shift and Modulation Data'. *Biophys. J.* **46**, 463.

Lakowicz, J. R., Gratton, E., Cherek, H., Miliwal, B. B. & Laczko, G. (1984b). 'Determination of Time-Resolved Fluorescence Spectra and Anisotropies of a

Fluorophore-Protein Complex using Frequency-domain Phase-modulation Fluorimetry'. *J. Biol. Chem.* **259**, 10967.

Lakowicz, J. R., Laczko, G., Gryczynski, I. & Cherek, H. (1986). Measurement of Subnanosecond Anisotropy Decays of Protein Fluorescence using Frequency Domain Fluorimetry. *J. Biol. Chem.* **261**, 2240.

Letokhov, V. S. (1987). "Laser Picosecond Spectroscopy and Photochemistry of Biomolecules". Adam Hilger, Bristol.

Levy, D. H. (1984). *Sci. Am.* **250**, 68.

Moog, R. S., Ediger, M. D., Boxer, S. G. & Fayer, M. D. (1982). Viscosity Dependence of the Rotational Re-Orientation of Rhodamine B in Mono- and Polyalcohols. Picosecond Transient Grating Experiments. *J. Phys. Chem.* **86**, 4694.

O'Connor, D. V. & Phillips, D. (1984). In "Time-correlated Single-photon Counting". Academic, London.

Phillips, D. (1984). 'A little light relief' *In* "Proceedings of The Royal Institution" (G. Porter & D. Phillips, Eds), Vol. 56, p. 161. Science Reviews, London.

Phillips, D. (1988). 'Supersonic Jet Spectroscopy: all that clusters is not cold, or is it?'. *In* "Proceedings of The Royal Institution" (D. Phillips Ed.) Vol. 59, p. 57. Science Reviews, London.

Phillips, D., Drake, R. C., O'Connor, D. V. & Christensen, R. L. (1985). Time-Correlated Single-Photon Counting (TCSPC) using Laser Excitation. *Anal. Instrument.* **14**, 267.

Phillips, D., Moore, J. N. & Hester, R. E. (1986). Time-Resolved Resonance Raman Spectroscopy Applied to Anthraquinone Photochemistry. *J. Chem. Soc. Faraday II* **82**, 2093.

Phillips, L. A. & Levy, D. H. (1986). 'The rotationally resolved electronic spectrum of indole in the gas phase'. *J. Chem. Phys.* **85**, 1327.

Porter, G., Reid, E. S. & Tredwell, C. J. (1974). Time-Resolved Fluorescence in the Picosecond Region, *Chem. Phys. Lett.* **29**, 469.

Porter, G., Tredwell, C. J., Searle, G. F. W. & Barber, J. (1978). Picosecond Time Resolved Energy Transfer in *Porphyridium cruentum*. Part 1. In the Intact Alga. *Biochim. Biophys. Acta* **501**, 232.

Pultney, S. K. (1972). Single-Photon Detection and Timing: Experiment and Techniques. *Adv. Elect. Elect. Phys.* **31**, 39.

Regan, J. D. & Parrish, J. A. (1982). "The Science of Photomedicine". Plenum, New York.

Rizzo, T. R., Park, Y. D. & Levy, D. H. (1986a). Dispersed fluorescence of jet-cooled tryptophan: Excited state conformers and intramolecular exciplex formation. *J. Chem. Phys.* **85**, 6945.

Rizzo, T. R., Park, Y. D., Peteanu, L. A. & Levy, D. H. (1986b). The electronic spectrum of the amino acid tryptophan in the gas phase'. *J. Chem. Phys.* **84**, 2534.

Stubbs, C. D., Meech, S. R., Lee, A. G. & Phillips, D. (1985). 'Solvent relaxation in lipid bilayers with dansyl probes'. *Biochim. Biophys. Acta* **815**, 351.

Szabo, A. (1984). 'Theory of Fluorescence Depolarisation in Macromolecules and Membranes *J. Chem. Phys.* **81**, 150.

Tao, T. (1969). Time-Dependent Fluorescence Depolarisation and Brownian Diffusion Coefficents of Macromolecules'. *Biopolymers* **8**, 609.

Tralau, C. J., MacRobert, A. J., Coleridge-Smith, P. D., Barr, H. & Bown, S. G. (1987). 'Photodynamic therapy with phthalocyanine sensitization quantitave studies in a transplantable rat fibrosarcoma. *Br. Cancer J.* **55**, 389.

Ware, W. R. (1971). "Transient Luminescence Measurements in Creation and Detection of The Excited State" (A. A. Lamola, Ed.), Vol. 1A, p. 213. Dekker, New York.

West, M. A. (1986). "Flash and Laser Photolysis in Investigation of Rates and Mechanisms of Reactions, 6" (C. F. Bernasconi, Ed.), p. 391. Wiley-Interscience, New York.

Wilkinson, F., Willsher, C. J., Leicester, P. A., Barr, J. R. M. & Smith, M. J. C. (1986). Picosecond Diffuse Reflectance Laser Flash Photolysis. *J. Chem. Soc. Chem. Commun.* 1216.

Yamazaki, I., Mimuro, M., Murao, T., Yamazaki, T., Yoshihara, K. & Fujita, Y. (1984). Excitation energy transfer in the light harvesting antenna system of the red alga porphyridium coventium and the blue-green alga *Anacystis nidulans* : Analysis of time resolved fluorescence spectra. *Photochem. Photobiol.* **39**, 233.

Yguerabide, J. (1972). Nanosecond Fluorscence Spectroscopy, *Meth. Enzymol.* **26**, 498.

# 7
# Ultraviolet Radiation Dosimetry with Polysulphone Film

B. L. DIFFEY

*Regional Medical Physics Department*
*Dryburn Hospital*
*Durham DH1 5TW, UK*

Ultraviolet radiation (UVR) is generally measured with thermal or photon detectors, often used in conjunction with optical filters. A different yet complementary approach is the use of various photosensitive films as UVR dosimeters. The principle is to relate the degree of deterioration of the films, usually in terms of changes in their optical properties, to the incident UVR dose. The principal advantages of the film dosimeter are that it provides a simple means of integrating UVR exposure continuously and that it allows numerous sites, inaccessible to bulky and expensive instrumentation, to be compared simultaneously.

The most widely used photosensitive film is the polymer polysulphone whose structural unit is shown in Fig. 7.1. The polysulphone is used in the form of a film, generally 40 μm thick, and mounted in a cardboard holder with a central aperture. This constitutes a dosimeter, or film badge (Fig. 7.2).

## 7.1. Preparation of Polysulphone Film

To prepare the film, the polysulphone is first purified by re-precipitation: a chloroform solution of the polymer is added to a stirred volume of

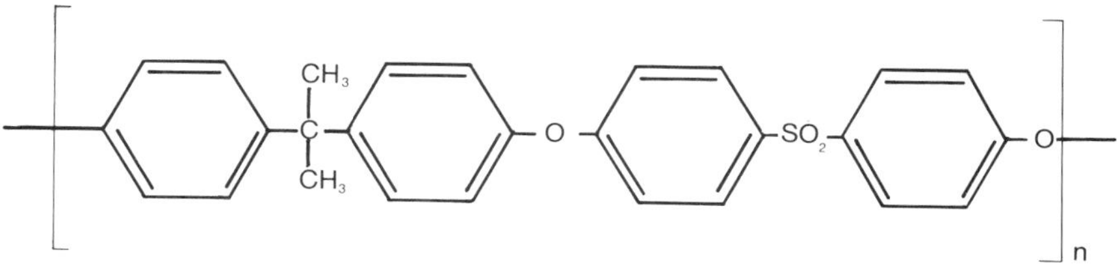

Fig. 7.1 The structural unit of polysulphone.

RADIATION MEASUREMENT IN PHOTOBIOLOGY
ISBN 0-12-215840-7

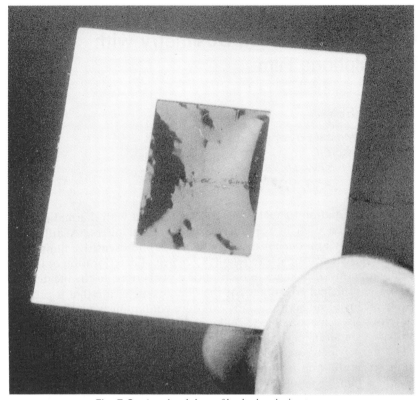

**Fig. 7.2**   A polysulphone film badge dosimeter.

methanol and then the filtered polymer is dried in a vacuum oven at $60°C$.
The film is made by spreading a 10% chloroform solution of re-precipitated
polymer on a flat glass plate with an adjustable casting blade, producing a
film between 36 and 44 $\mu$m thick. The film is removed from the glass plate,
dried overnight in a vacuum oven at $60°C$ and stored in the dark before use.
The choice of film thickness is a compromise between minimizing the
absorption of wavelengths greater than 330 nm and achieving mechanical
strength to facilitate handling (Davis *et al.*, 1976).

   If facilities to prepare the film are not available, it may be obtained either
in sheet form or mounted into individual film badge holders[a].

---

[a] Polysulphone film dosimeters are available from Dr A. Davis, 3 Cumley Road, Toothill,
Ongar, Essex CM5 9SJ, UK.

## 7.2. Optical Properties of Polysulphone Film

The absorption spectrum of 40 μm polysulphone film before and after exposure to UVR is shown in Fig. 7.3. The film may be used as a dosimeter for UVR by relating the incident radiant exposure (or dose) to the increase in absorbance measured at a wavelength of 330 nm (Fig. 7.4). The change in optical absorbance of the film at 330 nm ($\Delta A_{330}$) is determined by noting the absorbance of the film badge before and after irradiation in any standard UV spectrophotometer (Fig. 7.5). Unexposed polysulphone film of nominal thickness 40 μm has an optical absorbance at 330 nm in the range 0.15–0.19.

The spectral sensitivity of the film is confined principally to wavelengths less than 330 nm (Fig. 7.6). Since many photobiological processes, such as

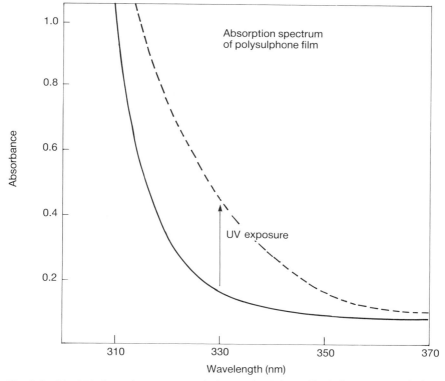

**Fig. 7.3** The UV absorption spectrum of 40 μm polysulphone film before (——) and after (------) exposure to ultraviolet radiation.

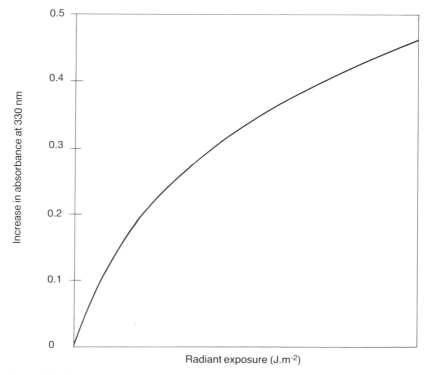

**Fig. 7.4**  General form of the dose-response curve for polysulphone film.

skin cancer and genetic damage in simple organisms, show similar spectral sensitivity, polysulphone film, if suitably calibrated, is appropriate as a "biological" dosimeter.

## 7.3.  Calibration of Polysulphone Film

The simplest approach to calibrating the film is to use "monochromatic" radiation from an irradiation monochromator and to plot the increase in absorbance at 330 nm ($\Delta A_{330}$) as a function of incident dose for a known wavelength and bandwidth combination on the monochromator. The form of the curve obtained is shown in Fig. 7.4. The incident dose (or radiant exposure), $H$, may be expressed in terms of the observed increase in optical absorbance at 330 nm ($\Delta A_{330}$) by the equation:

$$H = 1.2 \times 10^4 [\Delta A_{330} + (\Delta A_{330})^2 + 9(\Delta A_{330})^3]/S(\lambda) \ \text{J.m}^{-2} \qquad (1)$$

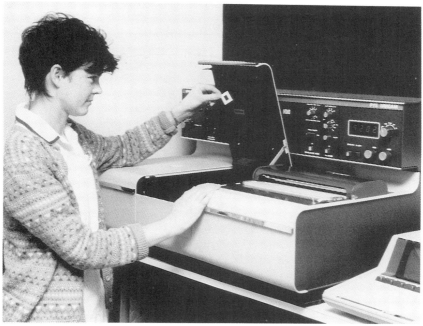

**Fig. 7.5**   Determining the absorbance of a polysulphone film badge in a spectrophotometer.

where $S(\lambda)$ is the spectral sensitivity of the film at wavelength $\lambda$ nm (Fig. 7.6).

For most applications in photobiology this may not be the most appropriate calibration since nearly all photobiological effects show a strong dependence on photon energy, or wavelength. For example, the radiant exposure of 320 nm radiation required to produce a given degree of erythema in human skin is about 100 times greater than the radiant exposure of 300 nm radiation needed to produce the same effect. In these situations where the interest lies in some particular action of the UVR, the effectiveness of the radiation is obtained by weighting the spectral irradiance according to the appropriate function of wavelength and then integrating over all wavelengths for which the spectral content of the source is non-zero. The determination of this single quantity, the biologically effective irradiance, is often the goal of photobiological ultraviolet dosimetry.

The biologically effective irradiance (UVR(BE)) may be expressed mathematically as:

$$UVR(BE) = \int E(\lambda)\varepsilon(\lambda)d\lambda \ \text{W.m}^{-2} \qquad (2)$$

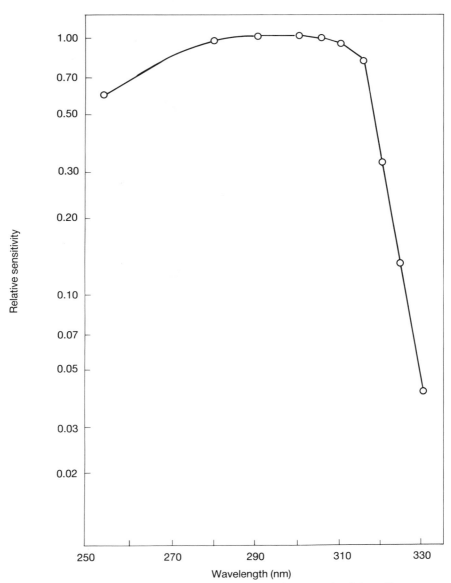

**Fig. 7.6** The relative spectral sensitivity curve of 40 μm polysulphone film.

where $E(\lambda)$ is the spectral irradiance (W.m$^{-2}$.nm$^{-1}$) at the site of interest and $\varepsilon(\lambda)$ is the relative effectiveness of the radiation at wavelength $\lambda$ nm in producing the desired biological effect (action spectrum) normalized to unity at some reference wavelength $\lambda_0$ nm. The quantity expressed by Equation (2) can be thought of as equivalent to that irradiance of monochromatic $\lambda_0$ nm radiation which would produce the same biological end point in a given time as the irradiance from the source in question, which is:

$$\int E(\lambda)\, d\lambda \; \text{W.m}^{-2}$$

The biologically effective radiant exposure, more commonly referred to as biologically effective dose, is simply the product of UVR(BE) and the time of exposure.

The determination of UVR(BE) may be achieved in two ways: either by measuring the spectral irradiance followed by numerical evaluation of the integral in Equation (2) or by direct measurement using a radiation detector whose sensitivity varies with wavelength according to the prescribed weighting function ($\varepsilon(\lambda)$).

### 7.3.1. Calibration by spectroradiometry

The former method is illustrated in Fig. 7.7. The spectral irradiance from a bank of ultraviolet fluorescent lamps is measured with a spectroradiometer (Optronic Model 742). Note that the polysulphone film badges are irradiated as close as possible to the input optics of the spectroradiometer. Once the spectral irradiance had been measured in 1 nm steps throughout the ultraviolet region, it was combined with an estimate of the erythema action spectrum ($\varepsilon(\lambda)$) given in Table 7.1. Logarithmic interpolation was used to estimate $\varepsilon(\lambda)$ at intermediate wavelengths. The biologically effective irradiance determined in this manner is equivalent to an irradiance of monochromatic 300 nm radiation which would result in the same degree of cutaneous erythema in the same exposure time as the irradiance from the array of lamps.

#### 7.3.1.1. Calibration using sunlight

The results of calibrating polysulphone film with natural sunlight are shown in Fig. 7.8. The technique used has been described in detail elsewhere (Diffey, 1987). Briefly, the spectral irradiance from 290 to 400 nm in steps of 1 nm was measured automatically on a flat, unshaded roof every half hour

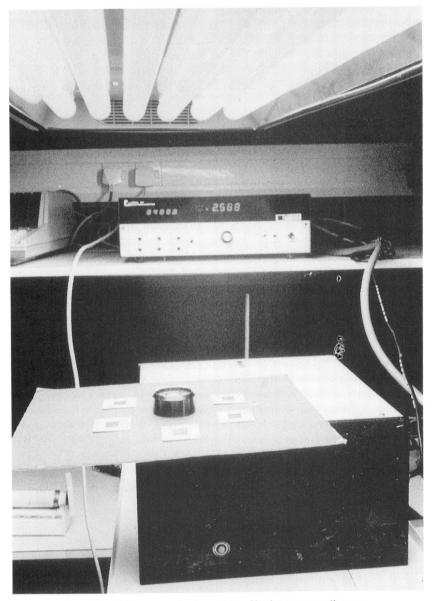

Fig. 7.7  Calibration of polysulphone film by spectroradiometry.

**Table 7.1** The erythema action spectrum (normalized to unity at 300 nm) used in the calculation of erythemally effective irradiance

| Wavelength (nm) | Relative erythemal effectiveness $\varepsilon(\lambda)$ |
|---|---|
| 250 | 1.74 |
| 260 | 1.90 |
| 270 | 1.43 |
| 280 | 1.31 |
| 290 | 1.43 |
| 295 | 1.51 |
| 300 | 1.00 |
| 305 | 0.51 |
| 310 | 0.075 |
| 320 | $8.7 \times 10^{-3}$ |
| 334 | $1.5 \times 10^{-3}$ |
| 365 | $5.7 \times 10^{-4}$ |
| 380 | $1.9 \times 10^{-4}$ |
| 405 | $1.4 \times 10^{-4}$ |

throughout one day in the summer and one day in the autumn. Fifty-five pairs of these dosimeters were exposed for all combinations of time interval given by $t_2 - t_1$ where $t_2$ ranged from 9.30 to 18.30 in hourly steps and $t_1$ ranged from 8.30 to 17.30 in hourly steps. In other words, a given pair of dosimeters would be exposed from $t_1$ until $t_2$. By this means it was possible to expose several pairs of dosimeters to approximately the same erythemally effective UV dose but for differing time periods.

The following day the optical absorbance of each dosimeter at 330 nm was determined in a spectrophotometer, and the difference ($\Delta A_{330}$) between this post-exposure absorbance and the absorbance determined prior to the exposure calculated for each dosimeter.

The erythemally effective dose received by a given film badge between the times $t_1$ and $t_2$ was estimated from the spectral irradiance determined every 30 minutes combined with the erythema action spectrum (Table 7.1). The observed $\Delta A_{330}$ for each film badge is plotted in Fig. 7.8. The solid line is an equation of the form:

$$\text{erythemally effective dose} = 2000[\Delta A_{330} + (\Delta A_{330})^2 + 9(\Delta A_{330})^3] \text{ J.m}^{-2}$$

(3)

For $\Delta A_{330}$ up to 0.3 (equivalent to an erythemally effective dose of about 1500 J.m$^{-2}$) analysis of variance indicates that the coefficient of variation

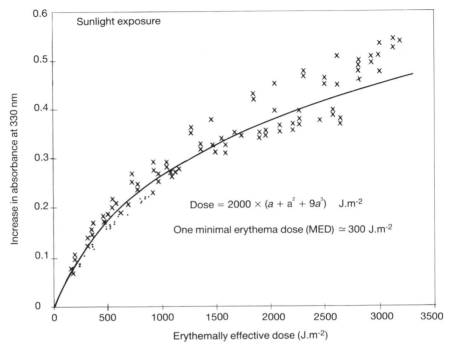

**Fig. 7.8** The increase in absorbance at 330 nm of polysulphone films plotted as a function of erythemally effective dose as a result of sunlight exposure for measurements obtained on a summer day (×) and an autumn day (•) at 55° N. The solid line is given by Equation (3).

(CV) on a dose estimated from a given $\Delta A_{330}$ is about 10%. However, it is clear that for $\Delta A_{330}$ in excess of 0.3 the polysulphone film becomes increasingly imprecise. Not only does the error on the estimated dose increase rapidly due to saturation (CV of ~30% for $\Delta A_{330} = 0.4$), but the variance of $\Delta A_{330}$ about the solid line becomes large. This is as a result of the fact that 40 μm thick polysulphone film has a spectral sensitivity extending up to about 330 nm (Fig. 7.6), whereas the erythema action spectrum shows a more rapid decrease in sensitivity with increasing wavelength in the interval 300–330 nm (Table 7.1). The consequence of this is that films exposed from early morning until late afternoon will receive an erythemally weighted dose which is not much higher than those films exposed from mid-morning until late afternoon; however the presence of radiation from 315 to 330 nm from early to mid-morning, which contributes very little to the erythemal dose, will cause some increase in absorbance in the polysulphone film.

## 7.3.1.2.  Calibration using artificial sources of UVR

When artificial sources of UVR are used as the radiation source of interest, the calibration accuracy is improved since the spectral emission remains much more stable than is the case with sunlight. Figure 7.9 shows the

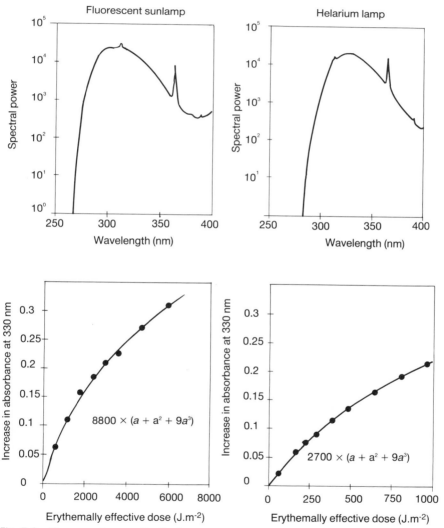

**Fig. 7.9**  The spectral power distribution (top) of two types of UV fluorescent lamp, and the corresponding dose-response curves (bottom) for polysulphone film.

spectral power distributions from two different ultraviolet fluorescent lamps, together with their respective dose-response curves. Note that the kernel of the dose-response equation contained within the brackets is the same for any UVR source (cf. Figs 7.8, 7.9 and Equation (1)). The only difference is the scaling factor which immediately precedes the term in the brackets.

### 7.3.2. Calibration by broadband radiometry

The alternative method of determining the biologically effective irradiance (UVR(BE)) is illustrated in Fig. 7.10. In this case, the sensor that is used has a spectral sensitivity closely approximating the "hazard curve" for occupational exposure to ultraviolet radiation (NIOSH, 1972). This type of calibration is useful when the polysulphone films are intended to be used as monitors for personal exposure to UVR in the workplace (see Section 7.5.2). It is vital, however, that calibration of the film badges is carried out using a light source with the same spectral characteristics as will be used in the proposed study since the spectral response of polysulphone film differs from the prescribed weighting function (see Fig. 7.11).

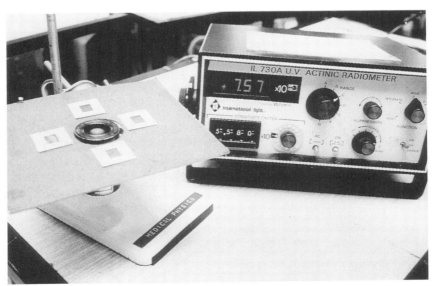

Fig. 7.10    Calibration of polysulphone film by broadband radiometry.

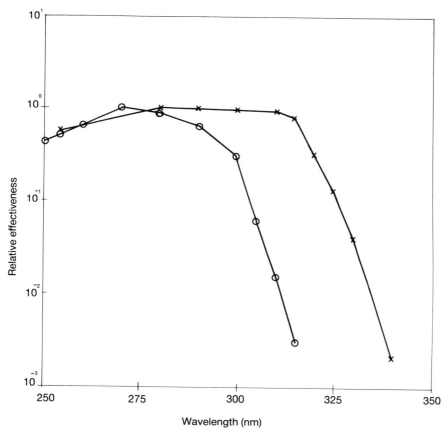

**Fig. 7.11**    The spectral sensitivity of 40 μm polysulphone film (×) and the NIOSH relative spectral effectiveness curve (○) for occupational exposure to UVR.

## 7.4.    Errors Associated with Polysulphone Film Dosimetry

There are two types of error associated with the calibration of any dosimetric system: systematic uncertainties in calibration and random uncertainties due to the reproducibility of the dosimeter. In this section only those factors which affect the reproducibility of response (random errors) of polysulphone film will be considered.

### 7.4.1.    Within-batch variation

Table 7.2 summarizes the results of exposing 10 polysulphone film badges

**Table 7.2**  Reproducibility of polysulphone film badges (after Davis and Gardiner, 1982)

| Film | $A_{330}$ Before exposure | After exposure[a] | $\Delta A_{330}$ |
|------|------|------|------|
| 1 | 0.170 | 0.465 | 0.295 |
| 2 | 0.169 | 0.459 | 0.290 |
| 3 | 0.166 | 0.461 | 0.295 |
| 4 | 0.170 | 0.459 | 0.289 |
| 5 | 0.168 | 0.455 | 0.287 |
| 6 | 0.170 | 0.461 | 0.291 |
| 7 | 0.171 | 0.461 | 0.290 |
| 8 | 0.172 | 0.460 | 0.288 |
| 9 | 0.169 | 0.453 | 0.284 |
| 10 | 0.175 | 0.474 | 0.299 |
|  |  | Coefficient of variation | 1.4% |

[a]30 minutes exposure in a "Climatest" artificial weathering chamber.

simultaneously for 30 minutes in an artificial weathering chamber (Davis and Gardiner, 1982). Exposing a number of films manufactured from the same batch under these controlled conditions yields a coefficient of variation of 1.4% Although many physical detectors, which generally measure dose rate rather than dose, might be expected to perform better than this the reproducibility is nevertheless adequate for many applications, particularly biological, where much larger variability associated with experimental design often has to be contended with.

### 7.4.2.   Dark reaction and effect of temperature

It was found (Davis *et al.*, 1976) that when stored, a previously exposed polysulphone film undergoes a "dark reaction". Table 7.3 illustrates that the $\Delta A_{330}$ measured immediately after exposure is about 8% less than that measured 24 h later and 10% less than that measured one week later (Kollias and Baqer, 1986). If exposed films are kept for several months their $\Delta A_{330}$ is about 5% higher than the values obtained 24 h after exposure (Diffey, 1987). It is important, therefore, that standardization of read-out time after exposure is adopted. The response of polysulphone film to ultraviolet radiation is unaffected by temperature during irradiation (Table 7.3).

Table 7.3 Effect of temperature and read-out time on polysulphone response (after Kollias and Baqer, 1986)

| Oven temperature (°C) during 30-min irradiation period (UVB) | Increase in absorbance at 330 nm read-out time after exposure | | |
|---|---|---|---|
| | Immediately | 24 h | 7 days |
| 25 | 0.176 | 0.190 | 0.192 |
| 37 | 0.179 | 0.197 | 0.199 |
| 52 | 0.181 | 0.200 | 0.205 |
| 70 | 0.184 | 0.197 | 0.200 |
| 85 | 0.181 | 0.193 | 0.194 |
| Mean ± 1 SD | 0.180 ± 0.003 | 0.195 ± 0.003 | 0.198 ± 0.005 |

## 7.4.3. Effect of surface contamination

The effects of surface contamination on the performance of polysulphone films have been studied by Tate (1979). A summary of these results is given in Table 7.4. Grease as a consequence of holding the film between the fingers, and dust shaken onto the film, both serve to increase the $\Delta A_{330}$ over "control" films. Removing surface contamination by cleaning with alcohol prior to read-out gives values which are in close agreement with those obtained from control films.

Table 7.4 The effect of surface contamination on the response of polysulphone films

| Badge group | $\Delta A_{330}$ |
|---|---|
| Control | 0.135 ± 0.005 |
| With grease | 0.143 ± 0.013 |
| With dust | 0.231 ± 0.038 |
| Grease-cleaned | 0.139 ± 0.006 |
| Dust-cleaned | 0.142 ± 0.007 |

## 7.4.4. The uncertainty in the measured dose

In Section 7.3 it was shown that the dose, or radiant exposure ($H$), may be derived from the observed increase in optical absorbance at 330 nm (denoted in this instance by $a$ rather than $\Delta A_{330}$) according to:

$$H = \text{source constant} \times (a + a^2 + 9a^3).\qquad(4)$$

By exposing a number of films to the same nominal dose, an estimate of the uncertainty in the average optical absorbance increase ($a$) may be obtained and expressed as $\Delta a$. The coefficient of variation of the dose derived from Equation (4) is then

$$\Delta H/H = 100 \times \Delta a(1 + 2a + 27a^2)/(a + a^2 + 9a^3)\%.\qquad(5)$$

Calculated values of $\Delta H/H$ are shown as a function of the mean increase in absorbance ($a$, or $\Delta A_{330}$) for three different estimates of $\Delta a$ in Fig. 7.12.

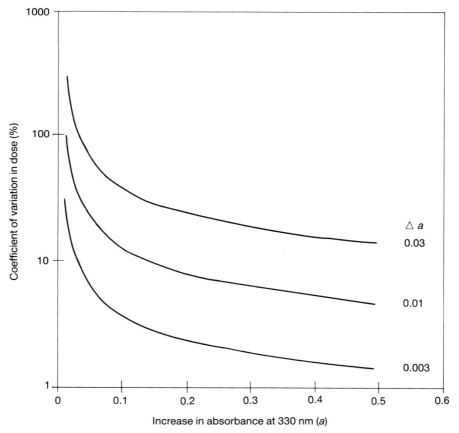

Fig. 7.12   The coefficient of variation in estimated dose plotted as a function of increase in absorbance at 330 nm ($a$) for three different values of the uncertainty ($\Delta a$) associated with $a$ (Equation (5)).

## 7.4.5. Increasing the reliability of dose measurements

Polysulphone film is normally mounted in a cardboard holder with a single rectangular aperture (Fig. 7.2). An alternative badge mount has been described (Gibbs *et al.*, 1984) incorporating four circular apertures (Fig. 7.13). Figure 7.13 illustrates that with the single aperture badge only a small fraction of the exposed surface area is used for spectrophotometry. By measuring absorbance at each of the four apertures, reliability of dose measurements increased by 30% compared with measurements made with a single-aperture badge (Gibbs *et al.*, 1984). The four-aperture badge mount has the same overall dimensions as the single-aperture badge, but because

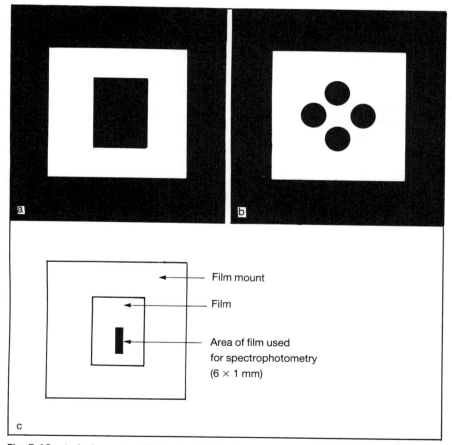

**Fig. 7.13** A single aperture film badge holder (a), a four-aperture film badge holder (b) and the area of polysulphone film used for spectrophotopmetry (c) (from Gibbs *et al.*, 1984).

the individual apertures are smaller, the former badges are more robust and afford more protection to the polysulphone film.


### 7.4.6. The problem of spectral sensitivity

A major criticism of systems such as polysulphone film for general photobiological dosimetry has focused on the mismatch between the spectral sensitivity of the film and biological action spectra (Calkins, 1982). However others have pointed out (Rupert, 1982) that because only limited kinds of spectral distributions and distribution changes occur, even with daylight illumination, it is possible that dosimetric errors arising as a result of mismatched action spectra could be partly corrected for.

By calibrating the film with the radiation source of interest using a spectroradiometer it is possible to derive functional relationships between the observed $\Delta A_{330}$ and biologically effective irradiance (or dose), as described in Section 7.3.1. Direct spectroradiometric calibration may not always be possible, however. If the relative spectral power distribution of the radiation source is known, a biologically effective dose can be derived from an observed $\Delta A_{330}$ by making use of the dose-response equation for monochromatic radiation (Equation (1)), the spectral sensitivity curve of polysulphone film ($S(\lambda)$; Fig. 7.6) and the action spectrum of the appropriate photobiological process ($\varepsilon(\lambda)$). These factors are related mathematically by Davis et al. (1976):

$$\text{biologically effective dose} = HQ \text{ J.m}^{-2} \qquad (6)$$

$H$ is the equivalent exposure (J.m$^{-2}$) of monochromatic 300 nm radiation which would result in the observed $\Delta A_{330}$ and $Q$ is a factor which takes into account the differences between the biological action spectrum and the spectral sensitivity of polysulphone film, and is defined as:

$$Q = \int P(\lambda)\varepsilon(\lambda)d\lambda / \int P(\lambda)S(\lambda) \, d\lambda \qquad (7)$$

where $P(\lambda)$ is the relative spectral power of the radiation source at wavelength $\lambda$ nm. The spectral sensitivity function $S(\lambda)$ is normalized to unity at 300 nm, and the biologically effective dose will be equivalent to a hypothetical exposure of monochromatic radiation at that wavelength at which $\varepsilon(\lambda)$ is normalized to unity.


## 7.5  Applications of Polysulphone Film Dosimetry

Polysulphone films have been used principally as personal dosimeters for evaluating human exposure to both natural and artificial UVR in a variety of situations. They have also been used as an alternative to physical detectors in monitoring radiation exposure in photochemical processes.

## 7.5.1. Personal exposure to natural ultraviolet radiation

It is evident that the natural UVR exposure received by different individuals will depend not only upon the quality and quantity of the UV environment but also on the behaviour of the individuals concerned. It might be expected

**Fig. 7.14** A subject wearing two polysulphone film badges: single aperture (left) and four aperture (right).

that outdoor workers, for example, would by and large receive much greater personal UVR doses than indoor workers. Nevertheless, it is difficult to estimate from the recording of stationary detectors the typical doses received by people under a variety of situations.

It is in the area of personal UVR monitoring, therefore, that polysulphone film dosimeters have been used most frequently. Film badges are normally worn on the lapel area (Fig. 7.14) as this site receives approximately the same UVR exposure as the face. The results of various field studies (Diffey *et al.*, 1982; Holman *et al.*, 1983; Larkö and Diffey, 1983; Webb, 1985; Schothorst *et al.*, 1985) indicate that indoor workers (excluding recreational exposure) receive about 2–4% of the annual ambient dose on a horizontal plane, and that outdoor workers receive annual doses which are some 3–5 times greater than those received by indoor workers. An indoor worker on a sun-seeking holiday may receive as much natural UVR during a 2-week summer vacation as he receives in the remaining 50 weeks of the year going about his normal duties.

### 7.5.1.1.  Anatomical distribution of sunlight

The small size of polysulphone film dosimeters means that they are particularly suited to measure the anatomical distribution of sunlight. Table 7.5 compares the mean fraction of ambient ultraviolet radiation received at different anatomical sites as measured on a rotating manikin (Diffey *et al.*, 1977) and living subjects pursuing a variety of outdoor activities such as hiking, boating and playing sports (Holman *et al.*, 1983). Inspection of Table 7.5 shows that for the cheek, hand and thigh the results obtained

**Table 7.5**  Comparison of mean fraction of ambient UVR received at anatomical sites of a rotating manikin and of living subjects

| Anatomical site | Manikin[a] | Living subjects[b] |
| --- | --- | --- |
| Cheek | 0.31 | 0.15–0.47 |
| Shoulder | 0.75 | 0.66–0.70 |
| Lower sternum | 0.66 | 0.44–0.46 |
| Lumbar spine | 0.47 | 0.58–0.71 |
| Upper arm | 0.52 | 0.59–0.66 |
| Dorsum of hand | 0.47 | 0.24–0.78 |
| Anterior thigh | 0.34 | 0.16–0.58 |

[a] Diffey *et al.* (1977).
[b] Holman *et al.* (1983).

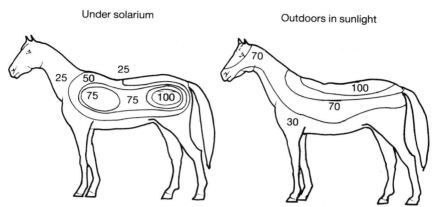

**Fig. 7.15**   The distribution of UVR on the surface of horses exposed to a UV solarium indoors and sunlight outdoors (from Keck *et al.*, 1987).

from the manikin were good approximations to the mid-points of the ranges found in living subjects. However, measurements of isolation on the manikin overestimated exposure of the lumbar spine and upper arm. These discrepancies may have resulted from a tendency of human subjects to stoop forward and outstretch the arm whilst manipulating objects, and a preference to turn away from the sun.

Measurements of the anatomical distribution of sunlight have not been confined to humans. Keck *et al.* (1987) used polysulphone film dosimeters to record the distribution of UVR at 17 sites on the surface of horses both outdoor and underneath UV solaria indoors. It was found that the anatomical distribution of UVR differed appreciably between these two exposure conditions (Fig. 7.15).

## 7.5.2. Occupational exposure to artificial UVR

Artificial sources of UVR are used increasingly in medical, industrial, military and consumer applications (Phillips, 1983). Examples where polysulphone films have been used to record UV exposure in the workplace are illustrated below.

### 7.5.2.1.   Industrial exposure

High-intensity sources of optical radiation find widespread application in industry. At one car factory in Sweden, fluorescent lamps are used as light

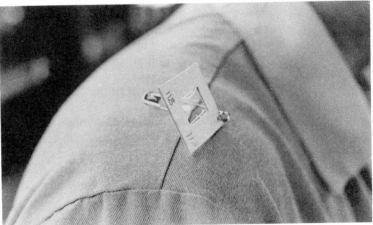

**Fig. 7.16** The paintwork inspection area in a car factory (top) and a subject wearing a polysulphone film badge on the shoulder (bottom) (from Diffey *et al.*, 1986).

sources to inspect the paintwork of newly fabricated cars (Fig. 7.16). Some of the people engaged in this process complained of rashes on the face and so personal monitoring to ultraviolet radiation exposure in the workplace was carried out with polysulphone film badges with a view to establishing whether or not the dermatological problems could be accounted for by excessive exposure to UVR. The results of this study (Diffey *et al.*, 1986) indicated very low UV exposure associated with this work practice with median values well below maximum permissible exposure (MPE) limits (NIOSH, 1972). In no case did the UV dose recorded on a polysulphone film badge exceed the MPE. Personal UV monitoring in this type of

situation can provide objective data which is welcomed by both management and the workforce.

## 7.5.2.2.  Hospital exposure

Ultraviolet radiation has a central role in the management of many
dermatological diseases. A variety of lamps are used in phototherapy and
the principal biologically effective emission may be either UVB or UVA
radiation.

Staff employed in departments where sources of UVR are used ther-
apeutically belong to that group of workers who are occupationally
exposed.

Polysulphone film dosimeters were used in one study (Larkö and Diffey,
1986) to determine whether nurses working primarily in "phototherapy
areas" received higher UV doses than nurses who worked in the non-
radiation areas of the dermatology department. The results, summarized in
Fig. 7.17, indicated that this was indeed the case, with something like 15%

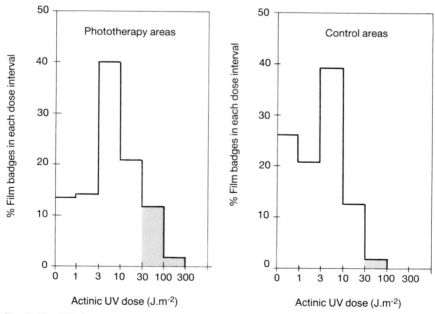

**Fig. 7.17**  UV dose recorded in phototherapy areas and non-radiation (control) areas of a
dermatology department. The shaded areas indicate those film badges which received doses in
excess of the 8 h maximum permissible exposure limit of 30 J.m$^{-2}$ (from Larkö and Diffey,
1986).

of the film badges recording doses in excess of the recommended 8 h maximum permissible exposure limit (NIOSH, 1972). Very rarely is the UV exposure of a member of staff sufficiently high to result in acute clinical symptoms such as erythema and keratitis, yet the results of personal UV monitoring can serve to remind staff to re-examine working practices so as to minimize the risk of long-term effects.

### 7.5.3.  Ambient solar UVR monitoring

Polysulphone films are sometimes used as an alternative to stationary physical detectors in long-term monitoring of ambient solar UVR (Davis *et al.*, 1979; Kollias and Baqer, 1986). Their principal advantages in this respect are that several dosimeters can be used simultaneously to record different facets of the solar environment (Qayyum and Davis, 1984), and that the inertness of polysulphone allows measurements to be made in environments which would be hostile to physical instrumentation without elaborate and expensive protection. An example of the second factor is given by Davis (1977) who measured the penetration of solar ultraviolet radiation into sea water using polysulphone films.

### 7.5.4.  Monitoring photochemical processes

Industrial photochemical processes include organic syntheses, photopolymerization, photografting, photocuring of surface coatings and photoresists (Phillips, 1983). In many of these applications polysulphone film may be appropriate for monitoring radiant exposure during sample irradiation. Davis and Gardiner (1982) have described the use of polysulphone film for assessing the deterioration of radiation output from lamps used in chambers to assess the weathering capability of polymeric materials.

## References

Calkins, J. (1982). Measuring devices and dosage units. *In* "The Role of Solar Ultraviolet Radiation in Marine Ecosystems" (J. Calkins, Ed.), pp. 169–179. Plenum, New York.

Davis, A. (1977). "Developments in Polymer Degradation". Applied Science Publishers, London.

Davis, A. & Gardiner, D. (1982). An ultraviolet radiation monitor for artificial weathering devices. *Polymer Degrad. Stab.* **4**, 145–157.

Davis, A., Deane, G. H. W. & Diffey, B. L. (1976). Possible dosimeter for ultraviolet radiation. *Nature* **261**, 169–170.

Davis, A., Howes, B. V., Ledbury, K. J. & Pearce, P. J. (1979). Measurement of solar ultraviolet radiation at a temperature and a tropical site using polysulphone film. *Polymer Degrad. Stab.* **1**, 121–132.

Diffey, B. L. (1987). A comparison of dosimeters used for solar ultraviolet radiometry. *Photochem. Photobiol.* **46**, 55–60.

Diffey, B. L., Kerwin, M. & Davis, A. (1977). The anatomical distribution of sunlight. *Br. J. Dermatol.* **97**, 407–410.

Diffey, B. L., Larkö, O. & Swanbeck, G. (1982). UV-B doses received during different outdoor activities and UV-B treatment of psoriasis. *Br. J. Dermatol.* **106**, 33–41.

Diffey, B. L., Larkö, O., Meding, B., Edeland, H. G. & Wester, U. (1986). Personal monitoring of exposure to ultraviolet radiation in the car manufacturing industry. *Ann. Occupat. Hygiene* **30**, 163–170.

Gibbs, N. K., Young, A. R. & Corbett, M. F. (1984). Personal solar UVR exposure: a method of increasing the reliability of measurements made with film badge dosimeters. *Photodermatology* **1**, 133–136.

Holman, C. D. J., Gibson, I. M., Stephenson, M. & Armstrong, B. K. (1983). Ultraviolet irradiation of human body sites in relation to occupation and outdoor activity: field studies using personal UVR dosimeters. *Clin. Exp. Dermatol.* **8**, 269–277.

Keck, G., Kasper, I., Schauberger, G. & Cabaj, A. (1987). The biological effect of UV irradiation of horses with artificial UV sources. *In* "Human Exposure to Ultraviolet Radiation: Risks and Regulations" (W. F. Passchier & B. F. M. Bosnjakovic, Eds), pp. 71–76. Elsevier, Amsterdam.

Kollias, N. & Baqer, A. H. (1986). Measurements of Solar Middle Ultraviolet Radiation in Kuwait. Environmental Protection Council, Kuwait.

Larkö, O. & Diffey, B. L. (1983) Natural UV-B radiation received by people with outdoor, indoor and mixed occupations and UV-B treatment of psoriasis. *Clin. Exp. Dermatol.* **8**, 279–285.

Larkö, O. & Diffey, B. L. (1986). Occupational exposure to ultraviolet radiation in dermatology departments. *Br. J. Dermatol.* **114**, 479–484.

National Institute for Occupational Safety and Health (1972). Criteria for a Recommended Standard ... Occupational Exposure to Ultraviolet Radiation. US Department of Health, Education and Welfare, Washington, DC.

Phillips, R. (1983). "Sources and Applications of Ultraviolet Radiation". Academic, London

Qayyum, M. M. & Davis, A. (1984). Ultraviolet radiation for various angles of exposure at Jeddah and its relation to the weathering of polyacetal. *Polymer Degrad. Stab.* **6**, 201–209.

Rupert, C. S. (1982). Photobiological dosimetry of environmental ultraviolet radiation. *In* "The Role of Solar Ultraviolet Radiation in Marine Ecosystems" (J. Calkins, Ed.), pp. 131–141. Plenum, New York.

Schothorst, A. A., Slaper, H., Schouten, R. & Suurmond, D. (1985). UVB dose in maintenance psoriasis phototherapy versus solar UVB exposure. *Photodermatology* **2**, 213–220.

Tate, T. J. (1979). Applications of Polymer films to ultraviolet radiation dosimetry. MSc Thesis, University of Kent, Canterbury, UK.

Webb, A. R. (1985). Solar ultraviolet radiation and vitamin D synthesis in man. PhD Thesis. University of Nottingham, UK.

# 8

# Computer Programs for Estimating Ultraviolet Radiation in Daylight

L. O. BJÖRN

*Department of Plant Physiology*
*University of Lund*
*Box 7007*
*S-220 07 Lund, Sweden*

## 8.1. Introduction

Measuring ultraviolet spectral irradiance of daylight requires expensive equipment and laborious calibration procedures. Estimating biological effects of the radiation requires repeated measurements at various times of the day and the year, and complex calculations. Alternatively, instead of measuring the ultraviolet spectral irradiance or the biologically effective ultraviolet dose rate, one could try to calculate it, and design computer programs which could also perform the integration or averaging over various time periods. Advantages are that computers have become cheap, and programs can be designed which are easy to use. A severe limitation is that computations cannot, even under optimal circumstances, give instantaneous values nearly as accurate as good measurements. On the other hand, measurements on a few occasions only may be misleading, as random fluctuations are large.

Data of solar spectral irradiance and total irradiance are required for many purposes, and not only in the ultraviolet. Technological use of solar power is one major field. For the design of heat-collecting systems spectral irradiance is less interesting than total irradiance and the main direction of the radiation, but for solar cell systems generating electric power the spectral distribution is important. The potential for photosynthetic biomass production is, as a first approximation, dependent on the number of photons between 400 and 700 nm. Specialists in materials, air pollution and meteorology have their special requirements. Because of this interest in solar radiation from scientists and engineers in various fields, the literature concerned with its estimation is scattered in many types of journals. I have tried in the present review to draw upon knowledge accumulated in the different fields.

RADIATION MEASUREMENT IN PHOTOBIOLOGY
ISBN 0–12–215840–7

We shall concentrate here on ultraviolet radiation and its biological effects, mainly the harmful ones. "Biological effects" in itself is a very wide topic, and we should remember that various biological effects have a rather different spectral dependence. Thus the action spectra for erythema, cancer initiation, inhibition of photosynthetic electron transport and inhibition of overall photosynthesis all differ.

Another thing to remember is that biological effects of UV radiation are in most cases dependent on the directional distribution of the radiation. Most models result in values for weighted or unweighted irradiance or irradiation on a horizontal surface, but with, for example, the model of Bird and Riordan (1986) one can compute the irradiance on tilted surfaces as well, and with, for example, the model of Björn and Murphy (1985) one can compute also weighted or unweighted fluence rate.

Two main approaches can be tried in trying to compute, or forecast, biological effects of daylight UV under various conditions. The first is the purely empirical approach (henceforth called the direct approach for brevity). We measure the biological effect in which we are interested for various times of the day, various days of the year, various locations, weather conditions, etc., and try to design a mathematical model which approximates our measured values as closely as possible, without considering the different wavelength components separately. The drawback of such an approach is that it is very inflexible: it is valid only for one effect and one organism, and if conditions change, e.g. by depletion of stratospheric ozone, the whole work has to be repeated.

The second approach relies more heavily on basic science. We measure biological effects in the laboratory under various radiation conditions (often a series of quasi-monochromatic radiations to obtain action spectra). Then we calculate the spectral distribution and direction of daylight by starting with the extraterrestrial solar spectrum and calculate the filtering and scattering effect of the atmosphere by taking into account what we know about the atmosphere and the apparent movement of the sun over the sky. Effects of ground reflectivity, transmission by water, canopies, sun lotions, etc. may also be taken into account. I shall refer to this as the spectral approach. This way has been tried by, for instance, Paltridge and Barton (1978) and Björn and Murphy (1985, relying heavily on Green, 1983).

In this approach usually direct (unscattered) light and diffuse (scattered) light are computed separately, and then added. It is, however, possible to compute the combined radiation directly (Gerstl et al., 1983 and literature cited by them).

Although computations in the spectral approach are very complex, this approach is feasible even with desk computers and has the advantages of flexibility and insight. Still, computation times for integration over time may be appreciable (hours). Laboratory measurements of action spectra may not be realistic enough. In particular, using "monochromatic" light one might overlook important synergistic or antagonistic effects of different wavebands. It is also difficult to deduce, for example, effects on a whole plant from measurements on a leaf, or effects on an active human subject from measurements on a small skin area in the laboratory. Therefore it is necessary to expose such theoretical estimates to critical tests. We cannot say that we are in the position to demonstrate such tests here. What we can, and will, do is to compare the results from two very different mathematical models, which correspond approximately to the empirical and the theoretical approaches. We shall start with the latter.

## 8.2. Extraterrestrial Solar Irradiance

In a point in space away from earth, solar radiation is essentially unidirectional, and its spectral irradiance, $H(\lambda)$, approximates that of black body radiation of 5269 K (Planck's radiation formula). The main deviation from the black body spectrum is due to Fraunhofer absorption lines of gases in the solar atmosphere, and their contribution to the overall spectrum is easily expressed in analytical form as "Gaussian modifiers" (Green, 1983).

Good experimental data for the UV part of the extraterrestrial solar spectrum (from satellite measurements) have been published by Heath and Park (1980). Green's (1983) analytical expression, based on these data, valid throughout the UVA and UVB regions, and corresponding to a bandwidth of 1 nm, has the form

$$H(\lambda) = 0.582(300 \text{ nm}/\lambda)^5[8972/\{\exp(2730.6 \text{ nm}/\lambda) - 1\}]$$

where six $A_i$s (four negative and two positive) and six $\sigma_i$s describe amplitudes and bandwidths of absorption and emission bands centred at six values of $\lambda_i$. The values of these constants are given in Table 2 of Green (1983).

Bird and Riordan (1986) have compiled values for the extraterrestrial spectrum for 122 wavelengths from 300 to 4000 nm corresponding to a bandwidth of 10 nm, using values from Neckel and Labs (1981) and Fröhlich and Wehrli (1981). Green and Chai (in press) have found the following analytical approximation for this range:

$$H(\lambda) = 2300[(728.5 \text{ nm}/\lambda)^{4.8}]/[\exp(728.5 \text{ nm}/\lambda)^2 - 1]$$
$$\{1 + \Sigma_i A_i \exp[-(\lambda - \lambda_i)^2/2\sigma_i^2] \text{ W.m}^{-2}$$

where the $A_i$s and $\sigma_i$s are new constants for ten new $\lambda_i$s.

Note that the latter formula, which is valid for the range 350–2600 nm, differs in form from the previous one, and is not of a Planck type.

Bird *et al.* (1983) discuss the sets of extraterrestrial spectral data of Thekaekara (1974), Labs and Neckel (1970) and Fröhlich (1980), and conclude that the latter is the most reliable. They calculate from it terrestrial spectra at atmospheric mass 1.5 for two reference directions (normal to the direction to the sun and tilted 37°).

## 8.3. Sun–Earth Distance

The above equations give solar spectral irradiance at the average distance of the earth from the sun. Since the distance between the two bodies varies during the year, $H(\lambda)$ should be multiplied by a factor, $f$, describing the inverse of the square of this variation, to give the spectral irradiance before radiation enters the earth's atmosphere. Spencer (1972) has given the following simplified equation, which gives $f$ with sufficient accuracy:

$$f = 1.000110 + 0.034221 \cos \theta_0 + 0.001280 \sin \theta_0 + 0.000719 \cos 2\theta_0$$
$$+ 0.000077 \sin 2\theta_0$$

where $\theta_0$ is (number of the day of the year, DN) $2\pi/365$.

Green and Chai (in press) have given a different approximation for the same purpose.

$$f = \{1 + 0.0167 \cos [(2\pi/365.25)(DN - 3.4)]\}^{-2}$$

and Josefsson (1986) uses the even simpler expression

$$f = 1 + 0.033 \cos (2\pi DN/365.25).$$

Since $f$ varies only by about 6.7% during the year, it is of little importance which approximation is used, and for most purposes the deviation of $f$ from unity can be neglected altogether.

## 8.4. Solar Elevation

Because the solar radiation is filtered through the atmosphere in different ways depending on the elevation of the sun above the horizon, it is necessary to know this angle. It can be computed with great precision, but for our purpose we get sufficient accuracy with very simplified computations. We have compared a combination of programs for astronomical purposes (Duffett-Smith, 1985: about 200 program lines, 7 kbyte) with a very simple program based on the following equations, and found that the simpler program is quite sufficient for the purpose of UV irradiance calculations.

$$DN = 30.3 \ (MO - 1) + DA$$

$$ED = 0.398 \sin \{(DN - 80)2\pi/365 + 0.0335[\sin(DN2\pi/365)] - \sin 1.3771\}$$

$$M = EDsin(LA\pi/180) + \cos(DI)\cos(LA\pi/180)\cos[(KL - 12)\pi/12]$$

$$\text{sun elevation} = ATN[M/SQR(1 - M^2)]180\pi \ \text{(degrees)}.$$

Here DA is date (number of day in the month), MO month number (1 for January, etc.), LA latitude (degrees, negative for southern latitude) and KL the hour of the day (official time, 24-hour clock). With just a small extension it is possible also to calculate the azimuth (compass direction) of the sun, which is important should one wish to compute the irradiance on a surface inclined to the horizontal.

One other program, accurate to $0.01°$, is provided by Walraven (1978).

Solar elevation affects radiation in two ways: (1) the lower the sun, the longer the path of the light through the atmosphere before it reaches the ground, and the greater the filtering effect of the atmosphere; (2) for the direct (unscattered) light from the sun, the direction determines how the light is distributed on the target. For a horizontal plane the irradiance due to direct radiation will be proportional to cos $Z$, where $Z$ is the zenith angle $(90° - \text{solar elevation})$. See below for irradiance on a tilted plane.

## 8.5. Effect of Cloudless Atmosphere

The atmosphere has two effects on the radiation from the sun: it absorbs and it scatters. This means that the radiation is not only weakened, but also that part of it becomes diffuse. To calculate total irradiance or fluence rate at ground level, it is usual to calculate the unscattered (direct) component and the scattered (diffuse) component separately, and then add them together.

In the UVB region the main absorber is ozone, but in urban atmospheres absorption by aerosol is also important. Both the atmospheric gases and aerosol contribute to scattering. All effects are wavelength dependent, but ozone absorption is more so than other processes. The humidity of the atmosphere also has an effect on the radiation climate; in the ultraviolet region mainly by its effect on aerosol scattering.

The inputs required for calculation are: column ozone (total ozone from ground level to top of atmosphere) in mm atm or Dobson units; humidity (relative humidity at ground level will suffice for approximate calculations);

air pressure at ground level (this is an expression for the amount of air that the radiation has to penetrate; approximate calculations can be carried out if the elevation above sea level is known, as pressure drops about 1 millibar per 8 m); a measure of aerosol or air turbidity. For the Green procedure the air should also be classified as urban, rural or marine.

We have investigated two sets of equations for computing the influence of cloudless atmosphere on radiation: those of Green (1983), which have been extended to longer wavelengths by Green and Chai (1988), and those of Bird and Riordan (1986). The latter model is limited to wavelengths greater than 300 nm. Another procedure, which we have not evaluated, is provided by Brine and Iqbal (1983). Yet another model which, however, is valid only for wavelengths greater than 400 nm, has been published by Goldberg and Klein (1980). These models include effects of solar elevation ($90°$ − zenith angle), scattering by atmospheric gases and aerosol, absorption by ozone and aerosol (and by water vapour in the infra-red), and reflection from the ground. We shall not recapitulate all details here: the interested reader is referred to Green (1983), literature cited therein, and to the program listing by Björn and Murphy (1985) given here as Appendix 1.

## 8.6. Estimation of the Amount of Ozone and Absorption by Ozone

We (Björn and Murphy, 1985) estimate column ozone from latitude, longitude and time in two steps. First the annual average for the location is determined from data by Gebhart et al. (1970). Then the deviation from the average for a particular time (day number, DA) is determined using analytical functions derived from satellite data over a 7-year period (Hilsenrath and Schlesinger, 1981). For $0-44°$ north latitude (LA) we used

$$\text{ozone (atm cm)} = \text{(annual average)}$$
$$+ 0.07[(LA + 10)/90]\cos\{[DN - 90 - (44 - LA)3.1]2\pi/365\}.$$

For $44-70°$ north latitude we used

$$\text{ozone} = \text{(annual average)} + 0.07[(LA + 10)/90]\cos[(DN - 90)2\pi/365].$$

Van Heuklon (1979) gives the analytic expression

$$\text{ozone(matm cm)} = 235 + \{150 + 40 \sin[0.9856(DN - 30)]$$
$$+ 20 \sin(3LO + I)\}(\sin^2 1.28LA)$$

where LA is latitude and LO longitude, and $I = 0°$ for the western and $20°$ for the eastern hemisphere. This expression certainly is not very accurate, and does not take modern measurements into account.

Long-term changes in ozone cannot be taken into account without certain assumptions. If one chooses to extrapolate the trend for the period 1979–1986 (Bowman, 1988), one should decrease the annual averages from Gebhart *et al.* (1970) by 0.7% per year for latitudes (LA) from 50° S to 20° N, and by (0.085LA − 1)% for latitudes from 20° N to 80° N.

The ozone optical density (naperian system, i.e. $\ln(I_0/I)$) for the sun in zenith is approximated by Green (1983) as amount of ozone in atm cm multiplied by the factor $11.277/(0.035 + \exp[(\lambda − 300)/7.150])$. For the wavelength interval 280–320 nm Cutchis (1974) uses the simpler expression $\exp(36.83 − 0.1151\lambda)$. $\lambda$ is the wavelength in nm in these expressions.

## 8.7. Other Effects of a Cloudless Atmosphere

Leckner (1978) and more recently Green (1983) and Bird and Riordan (1986) summarize the influence of water vapour, Rayleigh scattering (molecular scattering), aerosol scattering and aerosol absorption. These effects are more complicated to take into account than ozone absorption, and, of course, scattering gives rise to the diffuse component (skylight). For the variation in total (global) irradiance these effects are of less importance than the ozone absorption. The reader is referred to the cited work for details.

## 8.8. Reflection from the Ground or Water Surface

The ability of the ground to reflect radiation has an effect also on the amount of radiation striking organisms from above and from the sides. This is because part of the radiation reflected from the ground is scattered in a downward direction again by the atmosphere. We (Björn and Murphy, 1985) have incorporated in our computer program for UV estimates wavelength-dependent expressions for various ground covers which were proposed by Green (1983). For the UVB region this seems to be an unnecessary complication, since most natural ground covers have such a low reflectivity that reflection from the ground can be neglected. The exceptions are bright sand (UVB reflectivity $\approx 0.17$) and snow (UVB reflectivity 0.40–0.95 depending on age and structure; the older and the wetter the snow, the lower the reflectivity). A UV reflectivity of 0.8 means an increase in global UV radiation of about 25% (Josefsson, 1986). As mentioned by Josefsson (1986) it is a common misconception that water has a high UV reflectivity.

Forgan (1983) stresses that substituting albedo for UV reflectivity causes errors in the calculation of diffuse irradiance, especially with the sun close to zenith. The albedo is the ratio of reflected to incident radiant energy integrated over the whole spectrum.

## 8.9.  Cloud Effects

Presence of clouds can be only approximately accounted for. Higher accuracy can be obtained for long-time averages than for momentary values, since it is impossible to decide from ground observations what the thickness of the cloud cover is, and difficult to take into account the differences due to cloud type and whether the clouds obscure the sun or not. A complete cloud cover of the sky decreases, in a typical case, the biologically effective ultraviolet radiation by about 70%, and occasionally by more than 90% (Josefsson, 1986).

Cutchis (1980) uses, for mid-latitude sites, the simple, wavelength-independent correction factor $1 - 0.5C$, where $C$ is the "average cloud amount", which is available as a function of time of the year for some geographic sites, but insufficiently known for others. The relationship is obtained from a comparison of UV values measured with the Robertson–Berger meter (approximating erythemal radiation) and values computed by Cutchis' (1980) model.

Diffey (1984), based on data by Robertson (1972) uses the following three factors: clear sky 1, partial cloud 0.7, overcast sky 0.2.

Josefsson (1986), for calculating the UV radiation over a whole day, uses the correction factor $1 - 0.7C^{2.5}$, where $C$ is "total cloudiness". $C$ is obtained from three observations, at 0700, 1300 and 1900 hours local time, by counting the number of "octas" (eighths of the sky) covered by clouds at the three occasions, summing them and dividing by 24.

A way of computing the effect of cloudiness on radiation in the visible and near infra-red is described by Goldberg and Klein (1980). Here the input is tenths of the sky covered by cloud.

The effect of clouds on total radiation, and the interaction between clouds and ground albedo, are discussed by Kamada and Flocchini (1984).

## 8.10.  Direction of Reference Plane

When expressing irradiance one has to specify the direction of a reference plane. Usually when dealing with daylight, a horizontal reference plane is chosen, but for certain applications vertical planes or tilted planes are preferable.

Bird and Riordan (1986) discuss various methods for computing irradiances on non-horizontal planes and cite many other papers on the subject. They recommend the following formula, in which $\theta$ is the angle of incidence of the direct beam on the tilted surface (not zenith angle of the sun as in some other papers cited here), $t$ the tilt angle between the surface and the horizontal, $H(\lambda)$ the extraterrestrial solar irradiance at the mean earth–sun distance, $f$ the correction factor for the actual earth–sun distance, $Z$ the zenith angle of the sun, $I_d(\lambda)$ the direct (unscattered) irradiance on a horizontal surface, $I_s(\lambda)$ the scattered irradiance and $r_g(\lambda)$ the ground reflectivity:

$$I_T(\lambda, t) = I_d(\lambda)\cos\theta + I_s(\lambda) < \{I_d(\lambda)\cos\theta/[H(\lambda)f\cos Z]\}$$
$$+ 0.5[1 + \cos(t)][1 - I_d(\lambda)/(H(\lambda)f)] > + 0.5[I_d(\lambda)\cos Z + I_s(\lambda)]$$
$$r_g(\lambda)[1 - \cos(t)].$$

In some cases the above equation is impractical to use, since it requires determination of the incidence angle $\theta$. Suppose we know instead the compass direction (azimuth) of tilt, AZT. We may compute the sun's azimuth, AZ, by the following extension of the Björn and Murphy model (B&M):

2475.COS(AZ) = − (ED*COS(LA*PI/180) − SQR(1 − ED*ED)
*SIN(LA*PI/180)*COS(KL − 12)*PI/12))/SQR(1 − M*M)AZ =
ATN(COS(AZ)/SQR(1 − COS(AZ)*COS(AZ)) + PI/2

and then compute the cosine of the incidence angle as

COS($\theta$) = COS(Z)*SIN(t) + SIN(Z)*COS(t)*COS(AZ − AZT).

For many photobiological purposes, however, the spectral fluence rate rather than the spectral irradiance would be the proper way of expressing the spectral intensity of the radiation. The fluence rate refers to the radiation impinging on a sphere of unit cross section rather than that impinging on a unit area of plane surface. The spectral fluence rate can be obtained by changes in the models of Bird and Riordan (1986) or Björn and Murphy (1985). In the latter case a reasonable value can be obtained simply by omitting "M*" on line 3380 and changing S on the same line to 2*S. This procedure does not take into account direct reflection from the ground or surroundings, but the error is small. A slightly more accurate version of line 3380 for computing fluence rate would be

FLU(I) = H*(DIR + 2*(M*DIR*RG +
S*ML*EXP(− T1 − T2 − T3 − T4)/(1 − RA*RG)).

To give some idea of the relation between effective irradiance and effective fluence rate, a comparison was made between the two using the program of Björn and Murphy (1985) and weighing by the plant damage spectrum by Caldwell (A$\varepsilon$9, see Section 8.11). In one case with the sun 40° above the horizon the fluence rate was 85% higher than the irradiance; in another case with the sun 53° above the horizon the fluence rate was 68% higher. Theoretically, with completely isotropic radiation (coming equally from all directions) the fluence rate is four times as large as the irradiance; for light isotropic but limited to one side of the reference plane it is twice the irradiance; and for collimated light impinging perpendicularly on the reference plane they are equal.

## 8.11. Biological Action Spectra

### 8.11.1. Generalized medical effects spectrum

American Conference of Governmental Industrial Hygienists (ACGIH, 1978) have published human sensitivity values intended for the definition of safe working environments. It is an envelope curve for minimal spectral exposures causing erythema, photoconjunctivitis, etc. Wester (1981) has constructed analytic functions approximating these values:

230–270 nm $\quad$ ACGIH(1a) = $0.959^{270-\lambda}$
270–300 nm $\quad$ ACGIH(1b) = $1 - 0.36[(\lambda - 270)/20]^{1.64}$
300–315 nm $\quad$ ACGIH(1c) = $0.3 \; 0.74^{\lambda - 300}$
200–315 nm $\quad$ ACGIH(2) = $[161/(375 - \lambda)]15$
$\qquad\qquad\qquad\qquad$ $\exp\{ - [255/(375 - \lambda)]^{2.1}\}$.

Wester (1984) extends the validity of the last curve to 318 nm. Although with less precision than a combination of the second and third equations, it can thus be used for almost the whole UVB band.

### 8.11.2. Erythema

Parrish et al. (1982) have published values for erythema and melanogenesis action spectra extending to the visible. Their erythemal action spectrum is published as a table interpolated for every nm from 250 to 405 nm by Wester (1984). The present author found that the values above 280 nm can be analytically represented as

$$ERY1 = \exp( - 0.533 - 0.138x + 1.65 \times 10^{-4}x^3 + 6.69 \times 10^{-6}x^4 - 2.25 \times 10^{-7}x^5 + 2.31 \times 10^{-9}x^6 - 8.01 \times 10^{-12}x^7)$$

or, with higher accuracy

$$ERY2 = \exp(-0.4232 - 0.1413x - 0.0105x^2 + 2 \times 10^{-4}x^3 + 8.982$$
$$\times 10^{-6}x^4 - 3.921 \times 10^{-7}x^5 + 5.623 \times 10^{-9}x^6 - 3.603 \times$$
$$10^{-11}x^7 + 8.759 \times 10^{-14}x^8)$$

where $x = $ (wavelength in nm $- 300$).

An analytic expression for the erythemal action spectrum (based on Green and Mo, 1975, and older data) is given by Thimijan *et al.* (1978) as

$$ERY3 = 4 \exp[(\lambda - 296.5)/2.692]/\{1 + \exp[(\lambda - 296.5)/2.692]\}^2$$

where $\lambda$ stands for the wavelength in nm.

Diffey (1982) has critically reviewed six other investigations of the erythemal action spectrum, and stresses the variability, dependence on a number of factors, and the presence of two different erythemal processes. Therefore it is hardly meaningful to use very complicated analytical expressions in an attempt to achieve high accuracy. Nevertheless, to be able to compare different models appropriately, I constructed yet another expression approximating the action spectrum used in Diffey's (1984) computer program:

for $\lambda \leq 310$ nm, $ERY4a = 0.98 - 0.0957(\lambda - 300)$
for $\lambda > 310$ nm,

$$ERY4b = \exp[-5.0188 - 0.118(\lambda - 325) + 0.0009382(\lambda - 325)^2].$$

### 8.11.3. Generalized plant damage action spectrum and DNA spectrum

Caldwell (1968) presented a generalized plant damage action spectrum based on a variety of data. Thimijan *et al.* (1978), based on their own data on plant effects, judge the following function to be more appropriate:

$$A\varepsilon9 = [0.25(\lambda/228.178)^{9.0}]^4 \exp[4 - (\lambda/228.178)^{9.0}]$$

and this equation is also their best fit to Setlow's (1974) DNA action spectrum. Setlow's spectrum was originally intended to approximate the action spectrum for skin cancinogenests.

### 8.11.4. Photosynthesis

Rundel (1983), based on data of Caldwell, proposes several analytic functions. In later work Caldwell *et al*. (1986) refer to one of them:

$$PHO = 13.42 \exp(106.219 - 0.6122\lambda + 0.0008316\lambda^2).$$

The above is an action spectrum for inhibition of carbon dioxide fixation in a plant under irradiation with photosynthetically active light simultaneously with the inactivating UV. For inhibition of the capacity for photosynthetic electron transport by monochromatic UV without simultaneous photosynthetically active irradiation we (Bornman *et al*., 1984; Björn *et al*., 1986) and others have obtained much flatter action spectra.

## 8.12. Direct Approaches

As mentioned earlier, by direct approaches I mean methods that seek an estimate for the biologically effective UV exposure without considering each wavelength band separately. Such approaches have been taken by Cutchis (1980) and Josefsson (1986).

Cutchis (1980) tries to estimate the annual "damaging ultraviolet radiation" by finding formulae giving a quantity ($D$) agreeing with readings taken by the Robertson–Berger meter and correlating with incidence of skin cancer (Scotto *et al*., 1975). He starts out by assuming that $D$ can be decomposed into a product of factors, each of which is a function of one environmental variable only:

$$D = D_{ozone} D_{latitude} D_{altitude} D_{cloudiness} D_{albedo} D_{aerosol}.$$

For latitudes within 25° from the equator Cutchis arrives at

$$D = [1 - 0.00484(\tau - 240)](1 + 0.06h)\exp(-3.74 \times 10^{-4}L^2)$$
$$(1 - 0.50C)(1 + 0.50A)[1 - 0.093(\beta - 1)]$$

where $\tau$ = amount of ozone in Dobson units (matm cm), $h$ altitude in km, $L$ latitude in degrees, $C$ average cloud amount (1 for complete overcast), $A$ ground albedo and $\beta$ the amount of aerosol expressed in relation to a standard amount as defined by Green and Mo (1975).

For latitudes from 30° N to 55° N the corresponding formula for D is

$$(9.80 \times 10^{-6}\tau^2 - 0.010186\tau + 2.886)\langle 1 - \{9.08 \times 10^{-5}L^2 - 0.00528L$$
$$+ 0.0767 + (\tau - 256)[1.46 \times 10^{-4} + 1.34 \times 10^{-5}(L - 30^0) + 6.10 \times 10^{-7}$$
$$\delta(L - 45^0)^2]\}\rangle\exp(-3.74 \times 10^{-4}L^2)[1 + 0.010 + 1.20 \times 10^{-4}L^2$$
$$- 3.80 \times 10^{-3}L - (\tau - 197.8 - 2.46L)(1.424 \times 10^{-3} + 1.955 \times 10^{-6}L^2$$
$$- 9.25 \times 10^{-5}L)](1 + 0.06h)(1 - 0.50C)(1 + 0.50A)[1 - 0.093(\beta - 1)]$$

where $\delta = 0$ for latitudes from 30° N to 45° N and $\delta = 1$ for 45° N to 55° N.

A corresponding formula for D is derived also for mid-latitudes on the

Southern Hemisphere:

$$(9.80 \times 10^{-6}\tau^2 - 0.010186\tau + 2.886)[1 - 6.75 \times 10^{-5}L^2 + 3.75 \times$$
$$10^{-3}L - 0.0666]\exp(-3.74 \times 10^{-4}L^2)(1 - 9.08 \times 10^{-5}L^2 - 0.00528L$$
$$+ 0.0767 + (\tau - 256)[1.46 \times 10^{-4} + 1.34 \times 10^{-5}(L - 30°) + 6.10 \times 10^{-7}$$
$$\delta(L - 45°)^2])(1 + 0.06h)(1 - 0.50C)(1 + 0.50A)[1 - 0.093(\beta - 1)].$$

In using the formulae of Cutchis (1980), what has been said above about the use of albedo values in these contexts should be observed.

A more detailed model, although valid for a limited geographical region only, has been constructed by Josefsson (1986). His point of departure is the same factorial decomposition as that of Cutchis (1980). His model has the form

$$\text{DUV} = [1 + 0.033 \cos(2\pi DN/365.25)] f(\theta)\exp[-0.005134(OZ - 350)]$$
$$[1 + 0.08(1 - \text{beta}/0.05)][1 + 0.16(\text{fraction of time with snow}$$
$$\text{cover})][1 + 0.06h]$$
$$[(0.98 \text{ clear} + 0.85 \text{ broken} + 0.50 \text{ overcast})/\text{total}]$$

where

$f(\theta) = 48.782\sin(90° - \theta)$ mW.h.m$^{-2}$ for $90° - \theta$ (solar elevation) $> 15°$
and $f(\theta) = 0.0224 \times 10^{5.211 \sin(90 - \theta)}$ mW.h.m$^{-2}$ for $90° - \theta \leq 15°$,

OZ (amount of ozone in Dobson units, i.e. $\tau$ in Cutchis' symbolism) = $350 + (17.3 + LA)\sin[\pi(DN)/167]$ for DN $< 168$

and OZ $= 350 + (0.333 + 0.95LA)[\sin(\pi(DN + 30)/198]$ for DN $\geq 168$,

beta $= [31.0 + 1.0(59 - LA)]0.018\sin[2\pi(DN - 91)/365]$, and

clear, broken, overcast and total are the number of days with the respective type of sky, and the total days in the period over which the radiation is averaged.

The main factors of the expression for DUV describe the effects of the variation in (a) sun—earth distance as a function of daynumber DN of the year, (b) solar elevation ($h$ degrees), (c) Dobson units of ozone OZ for which a normal value can be expressed as a function of DN and latitude LA, (d) turbidity (aerosol), (e) surface reflectivity depending on fraction of the time with snow cover, (f) altitude ($H$ km), and the number of clear, broken and overcast days in relation to the total number of days in the period under consideration.

It should be noted that Cutchis' (1980) model estimates radiation effective for causing erythema and skin cancer, while Josefsson's model estimates damaging ultraviolet radiation defined by the ACGIH (1978) action spectrum as put in analytical form by Wester (1981) (see Section 8.11). These are not identical.

Josefsson's model also differs in allowing not only annual means but also monthly means.

## 8.13. Comparison Between Models

We can compare our (Björn and Murphy, 1985) spectral model for estimation of biologically effective ultraviolet radiation with the direct models of Cutchis (1980) and Josefsson (1986). However, the two direct models cannot be compared for two reasons. They base the calculations on different action spectra, and they are valid for different geographical regions. Already in this limitation a limitation of direct models becomes obvious.

The inputs for the different models are not exactly the same, so we must choose certain parameters to make a comparison. Josefsson's model (SMHI) calculates an aerosol value, while I have chosen aerosol = 0 for the Björn and Murphy (B&M) model. I had no cloudiness values for the B&M, so I also took SMHI values for cloudless conditions (the average reduction by clouds seems to be close to 30% the year around in Sweden for any whole month). SMHI uses average snow conditions as input, while I have used green farmland all year round for B&M, which is unrealistic for much of the year in Övertorneå close to the Arctic Circle. The barometric pressure was chosen as 1000 mb and the relative humidity (not important) 50%. Using SMHI, monthly average values were calculated in $mW.h.day^{-1}.m^{-2}$, while with B&M only a value for the fifteenth day of each month was calculated by adding weighted irradiance values for every hour. For both models the analytical function by Wester (1981) describing the ACGIH action spectrum was used for definition of sensitivity (this is easily changed in B&M but not in SMHI). We should have this in mind when judging the good agreement (Figs. 8.1 and 8.2) between the results obtained for the two models (I picked values of SMHI from the maps supplied by Josefsson, 1986). With linear regression of values for each month of the year for southern Sweden (Lund, $55.7°$ N, $13.4°$ E) and northern Sweden (Övertorneå, $66.5°$ N, $23.6°$ E), i.e. 24 pairs of values, the following relation was obtained:

$$DUV_{B\&M} = 1.031DUV_{SMHI} + 1.17, \ r^2 = 0.995 \text{ (cf. Figs 8.1 and 8.2).}$$

I also compared B&M and the program by Diffey (1984). In Diffey's (1984) program the spectral irradiance values are not normally visible to the user (the output is the number of minutes that one can stay in the sun without a sunburn, knowing the place, the time of start of exposure, the type of complexion, sunscreen and activity), but values were kindly provided by Dr Diffey. I made the comparisons of unweighted spectral irradiance at $20°$ and $60°$ N along the Greenwich meridian for summer times. The combined result for all wavelengths was

$$B\&M = 1.25\text{Diffey} + 0.0030 \text{ W.m}^{-2}.\text{nm}^{-1}, \ r^2 = 0.997$$

$$(B\&M/\text{Diffey} \pm SD = 1.23 \pm 0.21)$$

if I used the same ozone values for both programs and

$$B\&M = 1.22\text{Diffey} + 0.00016 \text{ W.m}^{-2}.\text{nm}^{-1}, \ r^2 = 0.998$$

$$(B\&M/\text{Diffey} \pm SD = 1.10 \pm 0.26)$$

if I let B&M generate its own ozone values.

I have also compared the results tabulated by Gerstl *et al.* (1983) with results obtained with the B&M program, and both have been compared also to the model of Bird and Riordan (1986). The input parameters for B&M

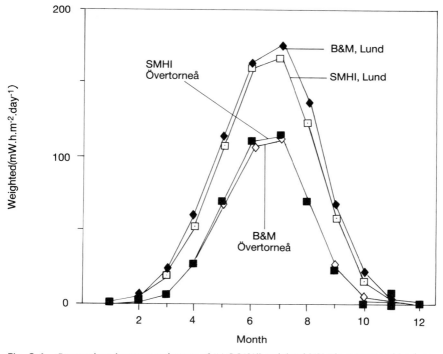

**Fig. 8.1** Comparison between estimates of "ACGIH"-weighted UV (time-averaged horizontal irradiance) by the B&M program and by the SMHI program. The SMHI values are averaged over each month considering normal cloudiness, while the B&M values were computed for the fifteenth of each month and a clear sky. Lund is in southern Sweden, Övertorneå in northern Sweden.

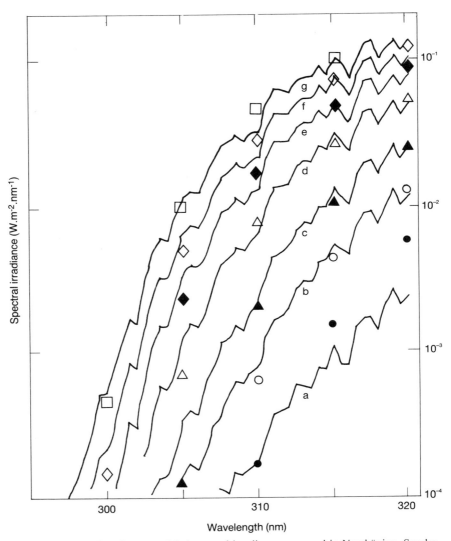

**Fig. 8.2** Comparison between global spectral irradiance measured in Norrköping, Sweden, on 18 Sept. 1985 (Josefsson, 1986, curves) and values generated by the B&M program (symbols) for various times of the day (h.min): (a) • 5.58; (b) ○ 6.43; (c) ▲ 7.15; (d) △ 8.05; (e) ◆ 8.55; (f) ◇ 9.46; (g) □ 11.53. When the sun was below the horizon (not shown), the model did not work. Zero aerosol was assumed.

were the same as described above for the comparison with SMHI, except that I used the same ozone values as tabulated by Gerstl *et al.* (1983), and $20°$ and $60°$ N, 1 January and 1 June, and 8.00, 10.00 and 12.00 hours (not all combinations). For the Bird and Riordan model (B&R) I chose no aerosol, 1 cm of precipitable water, tilt angle = 0, incidence angle = zenith angle and ground reflectivity zero. As measures of agreement I used the ratio between spectral irradiances obtained for the same wavelength (290, 300, 310 and 320 nm) for pairs of the models, and the correlation coefficient between the logarithms of the corresponding irradiances. The results are shown in Table 8.1.

As expected the agreement between the methods increases with wavelength. At 290 nm there is just barely agreement as to order of magnitude

**Table 8.1A**  Results of Björn and Murphy (1985) compared with those of Gerstl *et al.* (1983)

| Wavelength (nm) | B&M/Gerstl *et al.* ± SD | $r^2$ |
|---|---|---|
| 290 | 0.943 ± 0.962 | 0.89 |
| 300 | 0.546 ± 0.218 | 0.97 |
| 310 | 0.977 ± 0.124 | 0.99 |
| 320 | 0.996 ± 0.074 | 0.99 |

**Table 8.1B**  Results of Björn and Murphy (1985) compared with those of Bird and Riordan (1986)

| Wavelength (nm) | B&M/B&R ± SD | $r^2$ |
|---|---|---|
| 300 | 0.631 ± 0.205 | 0.99 |
| 310 | 1.086 ± 0.018 | 1.00 |
| 320 | 1.190 ± 0.045 | 0.99 |

**Table 8.1C**  Results of Bird and Riordan (1986) compared with those of Gerstl *et al.* (1983)

| Wavelength (nm) | B&R/Gerstl *et al.* ± SD | $r^2$ |
|---|---|---|
| 300 | 0.898 ± 0.351 | 0.99 |
| 310 | 0.902 ± 0.126 | 0.99 |
| 320 | 0.839 ± 0.075 | 0.99 |

between B&M and Gerstl *et al.*, while B&R, which is not particularly aimed at UV, does not extend to this low wavelength. At 300 nm B&M gives significantly lower values than both Gerstl *et al.* and B&R. At 310 and 320 nm the agreement is satisfactory between B&M and Gerstl *et al.*, while B&R gives lower values than the other two methods at 320 nm. Gerstl *et al.* have tabulated values for only two dates of the year, and B&R does not extend below 300 nm.

One spectral model constructed at the Solar Energy Research Institute (SERI), Golden, Colorado 80401 (similar to that published by Bird and Riordan) has been written in BASIC by L. Liedquist at the Swedish Testing Centre (SP, Box 857, S-501 15 Borås, Sweden), and a sample run was kindly put at my disposal by W. Josefsson, Swedish Meteorological and Hydrological Institute. A comparison between SERI and B&M computations is shown in Fig. 8.3. The B&M/SERI ratio averaged over wavelength from 300 to 400 nm was 1.12 ± 0.22 for global radiation and 1.14 ± 0.25 for diffuse radiation; the correlation coefficients between the logarithmic values are 1.00 in both cases. For this comparison turbidity was 0.1, precipitable water 1 cm and ground reflectivity 0.2 in the SERI model; aerosol 0.5,

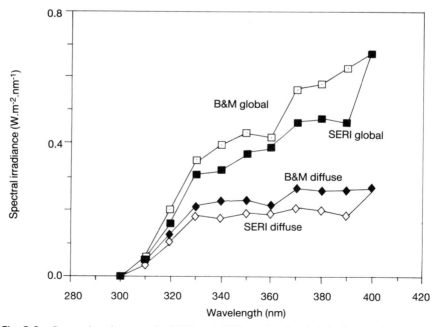

**Fig. 8.3** Comparison between the B&M and SERI models for global (direct + diffuse) and diffuse radiation.

relative humidity 0.5 and the ground green farmland in the B&M model. Both models simulated a day in late June with 0.3 atm cm of ozone and the sun $40°$ above the horizon.

Some spectral comparisons between the B&M predictions and spectro-radiometric measurements were shown by Björn and Murphy (1985). I have also made a comparison between the results recently published (Scotto *et al.*, 1988) for long-term (12 years) measurements with the Robertson–Berger meter on 8 locations in the United States and predictions with B&M. For this purpose the B&M program was loaded with the following analytic expression for the Robertson–Berger meter wavelength dependence.

$$R-B = \exp(-\ 0.03663 - 0.03808x - 1.625x^2 \times 10^{-3} - 3.716 \\ \times 10^{-6}x^3 - 2.013 \times 10^{-7}x^4 + 1.877 \times 10^{-8}x^5)$$

($x$ = wavelength in nm $-$ 300).

The comparison cannot be made in absolute units since the output of the R–B meter is counts rather than $W.h.m^{-2}$. Taking cloudiness into account by multiplying the program predictions by $[1 - 0.8*(\text{average sky cover})]$ the following relation was obtained:

$$R-B \text{ counts} = 0.134*B\&M(W.h.m^{-2}) + 7.02(r^2 = 0.942).$$

If cloudiness was disregarded the correlation coefficient was lower ($r^2 = 0.847$).

The inputs required for and the outputs obtained from the models mentioned above are as follows.

## B&M

Input: north latitude (we have not introduced an ozone algorithm for the Southern Hemisphere, but this can be done easily), month, date, time of day (hours on 24-h clock, local time), ozone amount (optional, the program can be asked to compute a normal value, or a value for some percentage depletion, in which longitude must also be provided), type of atmosphere (rural, urban, or maritime), ground cover, air pressure, relative humidity, aerosol level, wavelength interval and step size required (can give inter-polated values down to every nm), type of action spectrum (weighting function).

Output: wavelength, unweighted and weighted spectral irradiance on a horizontal surface. The program can integrate over wavelength, both unweighted and weighted spectra. Several action spectra (weighting functions) are built into the program.

**Table 8.2** Comparison of ozone values obtained or used in conjunction with different models. Values in Dobson units

|               | Diffey (1984) | Van Heuklon (1979) | B&M (1985) | Gerstl *et al.* (1983) |
|---------------|---------------|--------------------|------------|------------------------|
| Winter (DN = 1)   | 357 | 355 | 322 | 402 |
| Spring (DN = 90)  | 395 | 398 | 367 | —   |
| Summer (DN = 182) | 333 | 386 | 320 | 327 |
| Autumn (DN = 273) | 299 | 342 | 279 | —   |

**B&R**

Input: year, month, date, hours, minutes (local time), latitude, longitude, amount of ozone, precipitable water, turbidity (aerosol), tilt angle, incidence angle, wavelength interval required (gives values for every fifth nm from 300 nm up).

Output: wavelength, irradiance on a tilted plane (direct and diffuse can be obtained separately). The program does not integrate over wavelength in the form I have written the program, but this can easily be added.

**Gerstl *et al.***

Values have been looked up in tables where the entries are: one of two dates (1 January or 1 July), latitude, solar zenith angle (every $2°$ to $76°$), percentage ozone depletion and wavelength (every 5 nm from 290 to 320 nm). Diagrams are available for determining the zenith angle as a function of time and latitude.

Finally a comparison has been carried out between ozone values obtained with or used in conjunction with different models. I chose $50°$ N, $0°$ E for the comparison (Table 8.2).

## 8.14.  Activity and Orientation of Organisms

The activity of a mobile organism, as well as the growth habit of a plant, affects the effective UV exposure. In the case of cultivated plants, one might have effects of plant spacing as well as of the orientation of the rows, as has been observed for red/far-red effects (Kasperbauer and Karlen, 1986; Kasperbauer and Hunt, 1987). For human subjects, Diffey (1984) proposes the following exposure correction factors for different activities: sunbathing

1.0, exercising in sun 0.5, walking in partial shade 0.2, mountain skiing 0.8 (the latter factor includes the snow reflectivity effect).

## Acknowledgements

Thanks are due to Maria Rasmusson for help with computer programming, Weine Josefsson of the Swedish Meteorological and Hydrological Institute and Leif Liedquist of the National Swedish Testing Institute for making material available.

## References

ACGIH (1978). Threshold Limit Values for Chemical Substances and Physical Agents in the Workroom Environment with Intended Changes for 1978. American Conference of Governmental Industrial Hygienists.

Bird, R. E. & Riordan, C. (1986). Simple solar spectral model for direct and diffuse irradiance on horizontal and tilted planes at the earth's surface for cloudless atmospheres. *J. Climate Appl. Meteorol.* **25**, 87–97.

Bird, R. E., Hulstrom, R. L. & Lewis, L. J. (1983). Terrestial solar spectral data sets. *Solar Energy* **30**, 563–573.

Björn, L. O. & Murphy, T. M. (1985). Computer calculation of solar ultraviolet radiation at ground level. *Physiol. Vég.* **23**, 555–561.

Björn, L. O., Bornman, J. F. & Olsson, E. (1986). *In* "Stratospheric Ozone Reduction, Solar Ultraviolet Radiation and Plant Life" (R. C. Worrest & M. M. Caldwell, Eds), pp. 185–197. Springer, Berlin.

Bornman, J. F., Björn, L. O. & Åkerlund, H.-E. (1984). Action spectrum for inhibition by UV radiation of photosystem II activity in spinach [*Spinacea oleracea*] thylakoids. *Photobiochem. Photobiophys* **8**, 305–314.

Bowman, K. P. (1988). Global trends in total ozone. *Science* **239**, 48–50.

Brine, D. T. & Iqbal, M. (1983). Diffuse and global spectral irradiance under cloudless skies. *Solar Energy* **30**, 447–453.

Caldwell, M. M. (1968). Solar ultraviolet radiation as an ecological factor in alpine plants. *Ecol. Monogr.* **38**, 243–268.

Caldwell, M. M., Camp, L. B., Warner, C. W. & Flint, S. D. (1986). Action spectra and their key role in assessing biological consequences of solar UV-B radiation change. *In* "Stratospheric Ozone Reduction, Solar Ultraviolet Radiation and Plant Life" (R. C. Worrest & M. M. Caldwell), pp. 87–111, Springer, Berlin.

Cutchis, P. (1974). Stratospheric Ozone Depletion and Solar Ultraviolet Radiation on Earth. *Science* **184**, 13–19.

Cutchis, P. (1980). A Formula for Comparing Annual Damaging Ultraviolet (DUV) Radiation Doses at Tropical and Mid-latitude Sites. Report No. FAA-EE 80-21, US Department of Transportation, Administration, Office of Environment and Energy, Washington, DC.

Diffey, B. L. (1982). The consistency of studies of ultraviolet erythema in normal human skin. *Phys. Med. Biol.* **27**, 715–720.

Diffey, B. L. (1984). Using a microcomputer program to avoid sunburn. *Photodermatology* 1, 45–51.

Duffet-Smith, P. (1985). "Astronomy with your Personal Computer". Cambridge University Press, Cambridge.

Forgan, B. W. (1983). Errors resulting from the use of measured albedos to calculate diffuse irradiance. *Solar Energy* 31, 105–112.

Fröhlich, C. (1980). Photometry and Solar Radiation. Presented at the Annual Meeting of the Schweiz. Ges. f. Astrophysik und Astronomie.

Fröhlich, C. & Wehrli (1981). World Radiation Centre, Davos, Switzerland, magnetic tape.

Gebhart, R., Bojkov, R., & London, J. (1970). Stratospheric ozone: a comparison of observed and computed models. *Contrib. Atmos. Phys.* 43, 209–316.

Gerstl, S. A. W., Zardecki, A. & Wiser, H. L. (1983). UV-B Handbook, Vol. 1 Document No. LA-UR-83-728, Los Alamos National Laboratory, NM.

Goldberg, B. & Klein, W. H. (1980). A model for determining the spectral quality of daylight on a horizontal surface at any geographical location. *Solar Energy*, 24, 351–357.

Green, A. E. S. (1983). The penetration of ultraviolet radiation to the ground. *Physiol. Plant.* 58, 351–359.

Green, A. E. S. & Chai, S.-T. (1988). Solar spectral irradiance in the visible and infrared regions. *Photochem. Photobiol.* 48, 477–486.

Green, A. E. S. & Mo, T. (1975). Erythema Radiation Doses. CIAP Monograph 5, Part 1, Chapter 2, Appendix 1. Department of Transportation Climatic Assessment Program.

Heath, D. F. & Park, H. W. (1980). Ultraviolet Extraterrestrial Solar Spectral Irradiance, Geophys. Union Meeting, Toronto.

Hilsenrath, E. & Schlesinger, B. M. (1981). Total ozone seasonal and interannual variation derived from the 7 year Nimbus-4 BUV data set. *J. Geophys. Res.* 86, 12087–12096.

Josefsson, W. (1986). Solar ultraviolet radiation in Sweden. SMHI Reports: Meteorology and Climatology 53, 71 pp. Swedish Meteorological and Hydrological Institute, (S-60176 Norrköping, Sweden), Oct. 1986.

Kamada, R. F. & Flocchini, R. G. (1984). A general cloud transmittance modifier. *Solar Energy* 33, 631–632.

Kasperbauer, M. C. & Karlen, D. L. (1986). Light-mediated bioregulation of tillering and photosynthate partitioning in wheat. *Physiol. Plant.* 66, 159–163.

Kasperbauer, M. J. & Hunt, P. G. (1987). Phytochrome regulation of crop plant development under field conditions. Abstract 2-108-6, XIV Int. Botanical Cong., West Berlin.

Labs, D. & Neckel, H. (1970). Transformation of the absolute solar radiation data into the "International Practical Temperature Scale of 1968". *Solar Phys.* 15, 79–87.

Leckner, B. (1978). The spectral distribution of solar radiation at the earth's surface— elements of a model. *Solar Energy* 20, 143–150.

Neckel, H. and Labs, D. (1981). Improved Data of Solar Spectral Irradiance from 0.33 to 1.25 μ. *Solar Phys.* 74, 231–249.

Paltridge, G. W. & Barton, I. J. (1978). Division of Atmospheric Physics. Technical Paper no. 33. Commonwealth Scientific and Industrial Research Organization, Australia.

Parrish, J. A., Jaenicke, K. F. & Anderson, R. R. (1982). Erythema and melanogenesis action spectra of normal human skin. *Photochem. Photobiol.* **36**, 187–191.

Robertson, D. F. (1972). Solar Ultraviolet Radiation in Relation to Human Sunburn and Skin Cancer. PhD Thesis, University of Queensland.

Rundel, R. D. (1983). Action spectra and estimation of biologically effective UV. *Physiol. Plant.* **58**, 360–366.

Scotto, J., Fears, T. R. & Gori, G. B. (1975). Measurements of Ultraviolet Radiation in the United States and Comparison with Skin Cancer Data. National Cancer Institute, DHEW No. (NIH) 76–1029.

Scotto, J., Cotton, G., Urbach, F., Berger, D. & Fears, T. (1988). Biologically effective ultraviolet radiation: Surface measurements in the United States, 1974 to 1985. *Science* **239**, 762–764.

Setlow, R. B. (1974). Wavelengths in sunlight effective in producing skin cancer: A theoretical analysis. *Proc. Acad. Natl. Sci. USA* **71**, 3363–3366.

Spencer, J. W. (1972). Computer Estimation of Direct Solar Radiation on Clear Days, *Solar Energy* **13**, 437–438.

Thekaekara, M. P. (1974). Extraterrestrial solar spectrum, 3000-6100Å at 1-nm intervals. *Appl. Opt.* **3**, 518–522.

Thimijan, R. W., Carns, H. R. & Campbell, L. E. (1978). Final Report: Radiation Sources and Related Environmental Control for Biological and Climatic Effects UV Research (BACER). EPA-IAG-D6-0168, United States Environmental Protection Agency, Washington, DC.

Van Heuklon, Th. K. (1979). Estimating atmospheric ozone for solar radiation models. *Solar Energy* **22**, 63–68.

Walraven, R. (1978). Calculating the position of the sun. *Solar Energy* **20**, 393–397.

Wester, U. (1981). A Simple Formulae Approximation of the ACGIH Curve of Relative Spectral Effectiveness of Actinic UV. Internal Report RI 1981-02, Department of Radiation Physics, Karolinska Institute, Sweden.

Wester, U. (1984). Solar Ultraviolet Radiation in Stockholm—Examples of Spectral Measurements and Influences of Measurement Error Parameters. Internal Report RI 1984–03, Department of Radiation Physics, Karolinska Institute, Sweden.

# Appendix

## Program listing for "Daylight" by Björn and Murphy, 1985 (B&M)

The following program was described by Björn and Murphy (1985) but no program listing was published. The program is intended for computation of unweighted global spectral irradiance and irradiance integrated over a spectral interval. By additions of simple loops it can be used also for integration of irradiance over time. One line was recently added to avoid erroneous computations if times with the sun below the horizon are included in such an integration:

3385: IF ATN(M/SQR(1 – M*M)) < 0 THEN G(I) = 0.

REM: THIS IS THE APRIL 1988 VERSION OF THE PROGRAM "DAYLIGHT"
REM: WRITTEN BY L.O. BJÖRN AND T.M. MURPHY FOR COMPUTATION OF DAYLIGHT UV

**REM**: IT IS DOCUMENTED IN PHYSIOL. VEG. 23:555–561 (1985) AND "RADIATION MEASUREMENT IN PHOTOBIOLOGY" (ED. B.M. DIFFEY), ACAD. PRESS 1989
**REM**: TO RUN THE PROGRAM YOU NEED A MACINTOSH COMPUTER AND A MICROSOFT BINARY BASIC INTERPRETER FOR IT (VERSION 2.00 OR LATER)
**REM**: THE GRAPHICS CAN EASILY BE GREATLY IMPROVED FOR MACINTOSH, BUT WE WISHED TO KEEP A FORMAT WHICH DOES NOT REQUIRE EXTENSIVE MODIF. FOR OTHER COMPUTERS
1000: **PRINT** "first step: CALCULATE HEIGHT OF SUN"
1003: PI = 3.141592
**PRINT** "ENTER NORTH LATITUDE, MONTH, DATE, AND TIME OF DAY (24 HR CLOCK)"
**INPUT** LA, MO, DA, KL
**LET** DN = 30.3*(MO − 1) + DA
1050: **LET** ED = .398***SIN**((DN − 80)*2*PI/365 + .0335*(**SIN**(DN*2*PI/365) − **SIN**(1.3771)))
**LET** DI = **ATN**(ED/**SQR**(1 − ED*ED))
**PRINT** "DECL = "DI*180/PI"DEGREES"
M = ED***SIN**(LA*PI/180) + **COS**(DI)***COS**(LA*PI/180)***COS**((KL − 12)*PI/12)
**PRINT** "HEIGHT OF SUN IS "**ATN**(M/**SQR**(1 − M*M))*180/PI "DEGREES"
2000: **PRINT** "second step: FIND OZONE CONCENTRATION"
**PRINT** "DO YOU HAVE A VALUE IN MIND? (Y,N)"
**INPUT** A$
**DIM** O(9,18)
2070: **FOR** I = 1 **TO** 18
**FOR** J = 1 **TO** 9
2090: **READ** O(J,I)
**NEXT** J: **NEXT** I
2110: **DATA** .24,.24,.26,.28,.29,.32,.33,.35,.36
2120: **DATA** .24,.24,.26,.27,.29,.32,.34,.36,.36
2130: **DATA** .24,.24,.25,.27,.30,.33,.36,.37,.37
2140: **DATA** .23,.23,.25,.28,.32,.35,.38,.38,.37
2150: **DATA** .23,.23,.26,.29,.33,.37,.39,.38,.37
2160: **DATA** .24,.24,.25,.28,.31,.36,.38,.37,.36
2170: **DATA** .24,.24,.25,.27,.29,.32,.35,.36,.36
2180: **DATA** .24,.24,.25,.26,.28,.31,.33,.36,.36
2190: **DATA** .23,.24,.25,.28,.31,.32,.33,.34,.35
2200: **DATA** .23,.24,.25,.28,.31,.32,.33,.34,.35
2210: **DATA** .24,.23,.24,.27,.29,.32,.33,.34,.35
2220: **DATA** .24,.24,.24,.26,.28,.30,.32,.34,.35
2230: **DATA** .25,.25,.25,.27,.28,.30,.32,.34,.34
2240: **DATA** .26,.26,.27,.27,.28,.31,.32,.33,.34
2250: **DATA** .26,.26,.27,.28,.29,.33,.34,.33,.34
2260: **DATA** .25,.25,.27,.28,.32,.38,.36,.34,.34
2270: **DATA** .25,.25,.27,.28,.32,.35,.36,.35,.35
2280: **DATA** .25,.25,.27,.28,.30,.32,.33,.35,.35
2282: **IF** A$ = "N" **THEN** 2290
**PRINT** "ENTER YOUR VALUE"
2286: **INPUT** W3: **GO TO** 2500

```
2290: PRINT "ENTER LONGITUDE (WEST LONGITUDE AS NEGATIVE
NUMBER)"
INPUT LO
2310: LET IA = INT((LA + 5)/10) + 1
2320: LET IO = INT((LO + 170)/20) + 1
LET O0 = O(IA, IO)
2340: IF LA > 44 THEN 2370
LET DM = 90 + (44 − LA)*3.1
GO TO 2380
2370: LET DM = 90
2380: LET AM = .07*(LA + 10)/90
LET W3 = O0 + AM*COS((DN − DM)*2*PI/365)
PRINT "OZONE CONC IS" W3 "atm cm"
2500: PRINT "third step: CORRECT FOR ENVIRONMENT"
PRINT "WHAT ENVIRONMENT TYPE?", "1. RURAL, 2. URBAN,
 3. MARITIME"
2610: INPUT Z1
IF Z1 = 3 THEN 2680
IF Z1 = 2 THEN 2660
2640: KT = .255:K = 1.962:Q = .345:L1 = .122:LO = 439:AZ = .069:BE = 1.31:
NU = 5
2650: GO TO 2690
2660: KT = .288:K = 2.758:Q = .471:L1 = .0827:L0 = 510:AZ = .363:BE = 1.59:
NU = .9
2670: GO TO 2690
2680:KT = .106:K = 3.393:Q = .435:L1 = 1.049:L0 = 734:AZ = .032:BE = 2.44:
NU = 5
2690: PRINT "WHAT GROUND COVER?" PRINT 1. "PINE FOREST   ",
"2. OPEN OCEAN A"
PRINT "3. OPEN OCEAN B ", "4. GREEN FARMLAND"
PRINT "5. BROWN FARMLAND", "6. DESERT SAND"
PRINT "7. BLACK LAVA ", "8. GYPSUM SAND A"
PRINT "9. GYPSUM SAND B ", "10. SNOW COVER"
2720: INPUT Z2
2730: IF Z2 = 10 THEN 2910
IF Z2 = 9 THEN 2900
IF Z2 = 8 THEN 2890
IF Z2 = 7 THEN 2880
IF Z2 = 6 THEN 2870
IF Z2 = 5 THEN 2860
IF Z2 = 4 THEN 2850
IF Z2 = 3 THEN 2840
IF Z2 = 2 THEN 2830
2820: AO = .0147:DE = .05308:BA = .9181: GO TO 2920
2830: AO = .0653:DE = .07922:BA = .6636: GO TO 2920
2840: AO = .0511:DE = .05511:BA = .6013: GO TO 2920
2850: AO = .0441:DE = − .3948:BA = 25.88: GO TO 2920
2860: AO = .0417:DE = − .2449:BA = 24.95: GO TO 2920
2870: AO = .0387:DE = .08812:BA = 2.046: GO TO 2920
2880: AO = .0186:DE = .03422:BA = 1.302: GO TO 2920
```

```
2890: AO = .195:DE = .02156:BA = 1.956: GO TO 2920
2900: AO = .194:DE = .03028:BA = 1.764: GO TO 2920
2910: AO = .289: DE = .01409: BA = .472
2920: PRINT "WHAT AIR PRESSURE? (MILLIBARS)"
INPUT P
PRINT "WHAT RELATIVE HUMIDITY? (RANGE 0.0-1.0)"
2950: INPUT RH
PRINT "WHAT AEROSOL LEVEL?"
2970: INPUT W2
3000: PRINT "Wavelength", "   Direct", "   Diffuse", "  Global":PRINT "",
"Hor. Spectr. Irradiance, Watts per meter square per nm"
3001: READ G1, G2, G3, G4, A1, A2, A3, A4
3002: READ B1, B2, B3, B4, A, KU, T
3003: READ MA, MB, PA, PB, QA, QB, VA, VB
3005: DATA .5346, .6077, 1.0, 0, .8041, 1.437, .2864, 2.662
3006: DATA .4424, .1, .2797, 3.7, 84.37, .6776, .0266
3007: DATA 1.389, .5626, 1.12, .878, .8244, .8404, .4166, .1728
DIM G(121): DIM FLU(121)
3010: FOR I = 0 TO 120
3025: L = 280 + I
H = 1 - .738*EXP(- (L - 279.5)^2/(2*2.96^2))
3040: H = H - .485*EXP(- (L - 286.1)^2/(2*1.57^2))
H = H - .243*EXP(- (L - 300.4)^2/(2*1.8^2))
3060: H = H + .192*EXP(- (L - 333.2)^2/(2*4.26^2))
H = H - .167*EXP(- (L - 358.5)^2/(2*2.01^2))
H = H + .097*EXP(- (L - 368)^2/(2*2.43^2))
3100: H = .582*((300/L)^5)*(EXP(9.102) - 1)*H/(EXP(9.102*300/L) - 1)
3110 T1 = 1.0456*(P/1013)*((300/L)^4)*EXP(.1462*(300/L)^2)
3140 K2 = KT*(1 + K*EXP(- RH^ - 3)/(1 - RH)^Q*EXP(- (L - 300)/
(L0*(1 + L*RH*(1 - RH)^ - Q))))
AL = AZ*((300/L)^NU)*EXP(- BE*EXP((RH - 1)/.147))
T2 = (1 - AL)*W2*K2
T3 = W3*10.89*1.0355/(.0355 + EXP((L - 300)/7.15))
T4 = AL*K2*W2
3190 TX = .0018
TY = .0003
TZ = .0074
3220 MX = SQR((M*M + TX)/(1 + TX))
MY = SQR((M*M + TY)/(1 + TY))
MZ = SQR((M*M + TZ)/(1 + TZ))
3250 F = 1/(1 + A*(T3 + T4)^KU)
3260 F3 = 1/(1 + A3*T3^QA*W3^VA)
F4 = 1/(1 + A4*T4)
F6 = 1/(1 + B3*T3^QB*W3^VB)
F8 = 1/(1 + B4*T4)
FI = SQR((1 + T)/(M*M + T)) - 1
ML = (A1*T1^MA*F3 + A2*T2^PA*(1 + A1*(T1^MA)*F3))*F4
3320 RA = (B1*(T1^MB)*F6 + B2*(T2^PB))*F8
3330 IF (L - 300)/DE > 10 THEN 3360
3331: IF (L - 300)/DE < - 10 THEN 3362
```

RG = AO∗(1 + BA)∗**EXP**((L − 300)/DE)/(**EXP**((L − 300)/DE) + BA)
3350 **GO TO** 3370
3360 RG = AO∗(1 + BA)∗**EXP**(1)
3361: **GO TO** 3370
3362: RG = AO
3370: S = (F + (1 − F)∗**EXP**(− T3∗FI))∗**EXP**(− FI∗(G1∗T1 + G2∗T2))
3372: DIR = **EXP**(− T1/MX − T2/MY − T3/MZ − T4/MY)
3373: **IF ATN** (M/**SQR**(1 − M∗M)) < 0 **THEN** H = 0
3380: G(I) = H∗(M∗DIR + S∗ML∗**EXP**(− T1 − T2 − T3 − T4))/(1 − RA∗RG)
3385: FLU(I) = H∗(DIR + (M∗DIR∗RG + 2∗S∗ML∗**EXP**(− T1 − T2 − T3 − T4))/
(1 − RA∗RG))
**IF** L = 10∗**INT**(L/10) **THEN PRINT** L"nm    "H∗M∗DIR,   G(I) − H∗M∗DIR,
G(I)
3400: **NEXT I: PRINT** "DO YOU WANT SPECTRAL FLUENCE RATE
VALUES (Y/N)?":
**INPUT** FONF$: **IF** FONF$ = "N" **THEN** 3500
3410: **PRINT** "Wavelength, nm", "Spectral fluence rate, W m^ − 2 nm^ − 1"
3420: **FOR** I = 0 **TO** 120:L = 280 + I
**IF** L = 10∗**INT**(L/10) **THEN PRINT** L"       "FLU(I)
**NEXT** I
3500: **PRINT** "DO YOU WANT HARD COPY OF GLOBAL SPECTRAL
IRRADIANCE? (Y/N)"
**INPUT** A$
**IF** A$ = "N" **THEN** 4000
**GOSUB** 5000
3620:    **LPRINT**"    LOG    GLOBAL    SPECTRAL    IRRADIANCE
(Wm^ − 2 nm^ − 1)"
3630: **GOSUB** 5100
3680: **FOR** J = IL − 280 **TO** IU − 280 **STEP** IS
**IF** G(J) = 0 **THEN** LG = − 100 :**GO TO** 3720
**LET** LG = **INT**(10∗(**LOG**(G(J))/**LOG**(10)) − .5)
3720: **LPRINT** J + 280;
**FOR** I = − 70 **TO** 0
**IF** LG < I **THEN** 3770
**LPRINT** "I";
**NEXT** I
3770: **LPRINT**"   "G(J)
**NEXT** J

4000: **PRINT** "Fourth step: CHOOSE AN ACTION SPECTRUM"
**PRINT** "1.GREEN AND MILLER (DNA)"
**PRINT** "2.GREEN AND MOE (ERYTHEMA)"
**PRINT** "3.CALDWELL (PLANT)"
**PRINT** "4.IMBRIE AND MURPHY (ATPase)"
**PRINT** "5.ACGIH"
**PRINT**  "6.DIFFEY (ERYTHEMA)"
4030: **INPUT** Z3: **IF** Z3 = 6 **THEN** HILIMIT = 120 **ELSE IF** Z3 = 5 **THEN**
HILIMIT = 35 **ELSE** HILIMIT = 60
**IF** Z3 = 6 **THEN** LOLIMIT = 5 **ELSE** LOLIMIT = 0
**DIM** AC(121), GA(121), FLA(121):GI = 0:FLI = 0

**PRINT** "WL, nm";**TAB**(7)"ACTION SPECTR";**TAB**(20)"WEIGHTED
SP.IRR";**TAB**(40)"WEIGHTED SP.FLUENCE RATE"
4050: **FOR** I = LOLIMIT **TO** HILIMIT
**IF** Z3 = 6 **THEN** 4127
**IF** Z3 = 5 **THEN** 4123
**IF** Z3 = 4 **THEN** 4120
**IF** Z3 = 3 **THEN** 4110
**IF** Z3 = 2 **THEN** 4100
4090: AC(I) = **EXP**(13.82*( − 1 + 1/(1 + **EXP**((I + 280 − 310)/9)))):**GO TO** 4130
4100: AC(I) = 4***EXP**((I + 280 − 296.5)/2.692)/(1 + **EXP**((I + 280 − 296.5)/2.692))
^2:**GO TO** 4130
4110: AC(I) = **EXP**( − (((265 − I − 280)/21)^2)): **GO TO** 4130
4120: AC(I) = .0047*(.4***EXP**( − (((10 − I)/15)^2)) + **EXP**( − ((( − 10 − I)/
120)^2)) − .3092):**GO TO** 4130
4123: **IF** I > 20 **THEN** 4124: AC(I) = 1 − .36*((I + 10)/20)^ 1.64: **GO TO** 4130
4124: AC(I) = .3*.74^ (I − 20): **GO TO** 4130
4125: **REM**: THIS LINE AND 4126 ARE ALTERNATIVE FORMULAE FOR
DIFFEY: IF I > 30 THEN 4126: AC(I) = .98 − .0957*(I − 20):GOTO 4130
4126: AC(I) = **EXP**( − 5.01878 − .11795*(I − 45) + .0009382*(I − 45)*
(I − 45)):**GO TO** 4130
4127: AC(I) = **EXP**( − 6.861 − 3.631*(I − 70)/100 − 13.106*(I − 70)*
(I − 70)*.0001 − 35.784*(I − 70)*(I − 70)*(I − 70)*.000001 + 217.015*
((I − 70)/100)^4 + 59.639*((I − 70)/100)^5 − 499*((I − 70)/100)^6)
4130 GA(I) = G(I)*AC(I):FLA(I) = FLU(I)*AC(I)
GI = GI + GA(I):FLI = FLI + FLA(I)
**IF** I = 10***INT**(I/10) **THEN PRINT** I + 280;AC(I);**TAB**(20);GA(I);
**TAB**(40);FLA(I)"W m^ − 2 nm ^ − 1"
**NEXT** I
**PRINT** "INTEGRAL OF ACTION IS "GI, FLI" W m^ − 2"
4500 **PRINT** "DO YOU WANT HARD COPY OF WEIGHTED SPECTRAL
IRRADIANCE? (Y/N)"
**INPUT** A$
4520 **IF** A$ = "N" **THEN** 4800
4530 **GOSUB** 5000
**LPRINT** "ACTION SPECTRUM NUMBER"Z3
**LPRINT**
**LPRINT** "     LOG ACTION-WEIGHTED IRRADIANCE"
4620 **GOSUB** 5100
4680 **FOR** J = IL − 280 **TO** IU − 280 **STEP** IS
**IF** GA(J) = 0 **THEN** LG = − 100:**GO TO** 4720
**LET** LG = **INT**(10*(**LOG**(GA(J))/**LOG**(10)) − .5)
4720 **LPRINT** J + 280;
**FOR** I = − 70 **TO** 0
**IF** LG < I **THEN** 4770
**LPRINT** "I";
**NEXT** I
4770 **LPRINT** "    "GA(J)
**NEXT** J
**LPRINT** "INTEGRAL OF WEIGHTED IRRADIANCE AND FLUENCE RATE
FROM 280 to 340 nm are "GI,FLI"W m^ − 2"

```
LPRINT
LPRINT
LPRINT

4800 STOP
5000 LPRINT
5010 LPRINT "LATITUDE"LA"N; MONTH"MO"; DAY"DA"; TIME"KL
LPRINT "LONGITUDE"LO";OZONE CONCENTRATION"W3
LPRINT "ENVIRONMENT TYPE "Z1"; GROUND COVER"Z2";
PRESSURE"P"MILLIBARS"
LPRINT "RELATIVE HUMIDITY"RH"; AEROSOL LEVEL"W2
LPRINT
5060 RETURN
5100 LPRINT "    − 7   − 6   − 5   − 4   − 3   − 2   − 1   − 0"
LPRINT "      |     |     |     |     |     |     |     |"
5140 PRINT "ENTER LOWER WAVELENGTH BOUND (MIN 280 nm)"
INPUT IL
PRINT "ENTER UPPER WAVELENGTH BOUND (MAX 340 nm)"
INPUT IU
PRINT "ENTER INCREMENT"
INPUT IS

RETURN
```

# 9
# Optical Radiation Interactions with Living Tissue

M. SEYFRIED

*Universität Karlsruhe*
*Botanisches Institut 1*
*Kaiserstr. 12*
*D-7500 Karlsruhe 1, FRG*

## 9.1. Introduction

In photobiological and photomedical research light is applied to biological tissue in order to induce an effect or a response. Such a response or a set of responses from different light treatments may be used to analyse the properties of photoreceptors and reaction chains involved in bringing about the responses. In medical applications light may also serve as a therapeutic agent.

In experimental photobiology light may also be employed as a measuring tool. Here light is used to quantify the presence or synthesis of pigments or other light-absorbing molecules, or to gain information on structure or structural changes of biological tissues. In medical terms this means that light is a diagnostic tool.

Obviously this differentiation creates two groups of experimental approaches to photobiology. In the first group we put all those experiments where simply the physical interaction of light with the tissue (or more generally, system of biological origin, which may also be a solution or suspension) is of interest and measured. These interactions are absorption, scattering and surface reflection, as will be specified later on. The second group includes experiments or medical procedures, where the physical interaction leads to the observable, or beneficial effects. The Grotthus–Draper law states that only absorbed light can be effective in triggering reaction chains. Therefore, the relevant interaction in the second group is absorption. Table 9.1 gives an overview of typical photobiological and photomedical phenomena and their grouping under the above-defined terms.

A few selected examples may serve to illustrate uses of light in photobiological research or photomedicine. Although I will use some technical terms that will be explained in more detail later on in this chapter, these

RADIATION MEASUREMENT IN PHOTOBIOLOGY
ISBN 0–12–215840–7

192

**Table 9.1**   Selected photomedical and photobiological phenomena

| Topic | Observed effect(s) | Problems associated | References |
|---|---|---|---|
| UV action on skin | Erythema, tanning, carcinogenesis | Penetration of light into multilayered structure, identity of photoreceptor, localization of photoreceptor, dynamic effects[a] | Harm (1980), Regan and Parrish (1982), Diffey (1983) |
| Photomorphogenesis | Enzyme induction, de-etiolation, phytochrome photoconversion, red–far-red reversibility | Penetration of light into a complex structure, light and dark kinetics of photoreceptor phytochrome, involvement of other photoreceptors, dynamic effects[a], light piping | Hartmann (1983), Schäfer et al. (1983), Seyfried and Schäfer (1983), Mandoli and Briggs (1982) |
| Phototropism | Phototropic bending of hypocotyl or coleoptile | Light distribution in a cylindrical organ; nature, mode of action and localization of photoreceptor | Parsons et al. (1984), Baskin and Iino (1987), Steinhardt et al. (1987) |
| Fluorescence | Fluorescence of natural and artificially introduced fluorochromes | Distribution of exciting light, propagation of fluorescent light, bleaching | Lork and Fukshansky (1985), Govindjee et al. (1986) |
| Photosynthesis | Induction of photosynthesis, steady state photosynthesis | Light distribution, energy transfer, adaptation, state transition, chloroplast movement, competing reactions | Terashima and Saeki (1983), Weis (1985), Gausman (1985), Govindjee et al. (1986) |
| Medical diagnostics | Reflectance from enamel, oxygen saturation in blood or muscle | Scattering properties of tissue, light distribution | Groenhuis et al. (1983), Lübbers and Wodick (1975) |

[a] Dynamic effects are changes of one or several of the relevant quantities with time: changes in the amount or localization of photoreceptors with time, changes in the amount of other pigments (dynamic screening), changes of the kinetics of the photoreaction chain with time.

terms may be understood quite intuitively without risking misinterpretation.

For a first example let us extract a pigment, say chlorophyll, from a tissue. It is then possible to determine the amount of the pigment extracted by calculating the concentration of pigment in a known amount of extraction buffer from the absorbance of the sample measured in a spectrophotometer against a suitable reference. This is a standard laboratory procedure, and provided we are really dealing with a clear pigment solution, this measuring procedure involves only one type of light–matter interaction—absorption. Light simply goes straight through the sample with a part of it being absorbed. What we can measure is the part of light not absorbed, and an easy calculation gives the pigment concentration.

Next we try to make things simpler by omitting the possibly cumbersome extraction step and measuring the pigment *in vivo* or *in situ*. Now the pigment in question is imbedded in a structure which itself interacts with a measuring light beam, either through scattering photons from the beam out of their original direction or through absorption by the other constituents of the tissue. In addition, our measurement suffers from the lack of a suitable reference which was easy to provide in the former example of a cuvette measurement.

Another complication arises when one becomes aware of the fact that the pigment of interest may be unevenly distributed along the path of the beam, e.g. in the case of chlorophyll in a leaf where a high concentration of chlorophyll is found in the palisade tissue, a lower concentration in the spongy tissue and none in normal epidermal cells. Pigment distribution can also be inhomogeneous over the cross section of a light beam. In a leaf epiderm, chlorophyll is only found in stomatal cells, so that a light beam through the epiderm can interact with chlorophyll-free parts and with chlorophyll-containing parts of the tissue. Looking even more closely, one will find (still in the chlorophyll example) that chlorophyll always comes in little packages, chloroplasts, while the vacuole that takes up most of the cell volume is chlorophyll free. This concludes the second example of light as a measuring tool where we have only very elementary physical interactions of light with the tissue, but where a complex sample geometry makes measurements and interpretation of measured data a difficult task.

Our next problem (example) is taken from photomedicine and introduces additional considerations. In photodynamic therapy (Dougherty *et al.*, 1982; see also Chapter 6) a photosensitive dye, haematoporphyrin, is administered and retained preferentially in malignant tissues. Irradiation of the tissue containing haematoporphyrin and absorption of light by haematoporphyrin results in formation of singlet oxygen with subsequent cell destruction. Obviously a major concern with this therapy is that the formation of singlet oxygen is undesirable in normal cells so that the therapeutic light should be provided at a rate sufficient to kill malignant cells (with a higher content of haematoporphyrin) but tolerable for normal cells. But how do we know how much light penetrates to a given place somewhere in the depths of the tissue? It is not surprising to learn that spatial light distribution depends on the already-mentioned light–tissue interactions, surface reflection, absorption and scattering. These interactions are wavelength dependent, so that there may be wavelengths where much light reaches the tumour tissue while at other wavelengths most of the light would be absorbed in the overlying tissue. As a second parameter in this dosimetry problem we have to consider the absorption properties of haematoporphyrin. In practice a wavelength of 632 nm is chosen because human tissue is relatively transparent for red light and also because a practicable light source (He–Ne laser) is available for this wavelength. As it turns out, haematoporphyrin has only a minor absorption peak in the red (the main absorption band is in the UVA to blue region), so that we have a typical compromise situation.

One last example will relate to *in vivo* action spectroscopy. In action spectroscopy (see Chapter 5) biological systems are subjected to a sequence of light irradiations at several wavelengths and the resulting light-dependent responses are monitored. The aim of action spectroscopy is to identify photoreceptor pigments, their location and pathways of photoreactions, or to collect information on the state of the photoreceptor system (cf. Hartmann, 1983). Instead of elaborating upon various examples of action spectroscopy which would easily exceed the space allocated for this chapter I refer the reader to two review articles on action spectroscopy by Hartmann (1983) and Schäfer *et al.* (1983) and the references given there, as well as to the action spectroscopy chapter in this book. Table 9.1 may also be read as a list of selected problems associated with *in vivo* action spectroscopy. It is evident that, apart from specific problems related to each of the cases, the central point is to know the integral photon fluence rate at the site of the (sometimes unknown) photoreceptor. In the table the expressions "light distribution" or "penetration of light" have been used; elsewhere "light gradient" (Seyfried and Fukshansky, 1983), "photon flux gradient" (Hartmann, 1983; Fukshansky-Kazarinova *et al.*, 1986), "local space irradiance" (Grum and Becherer 1979) and other terms serve to describe more or less the same state of affairs. In the following I will provide some basic physical facts that will later on be used to derive models of light propagation and distribution in biological tissues. A major stress will be put on practicability and usefulness in the laboratory routine.

## 9.2  Basic Light–Matter Interactions

### 9.2.1.  Surface reflection

At the interface between two media of different refractive indices, $n_0$ and $n_1$, part of an incident light beam is reflected, another part is refracted (bent). Figure 9.1 illustrates this for the case of $n_1 > n_0$; Snell's law describes the bending:

$$\frac{\sin \beta}{\sin \alpha} = \frac{n_0}{n_1}. \tag{1}$$

The Fresnel equation (Equation 2) can be used to quantify the reflection from the surface:

$$R_\alpha = \frac{1}{2} \left( \frac{\sin^2 (\alpha - \beta)}{\sin^2 (\alpha + \beta)} + \frac{\tan^2 (\alpha - \beta)}{\tan^2 (\alpha + \beta)} \right). \tag{2}$$

For vertical incidence, $\alpha = 0^0$, a simpler form applies

$$R_\perp = \left( \frac{n_1 - n_0}{n_1 + n_0} \right)^2. \tag{3}$$

For light emerging from the medium with higher refractive index all light incident beyond the critical angle

$$\beta_c = \sin^{-1} \left( \frac{n_0}{n_1} \right) \tag{4}$$

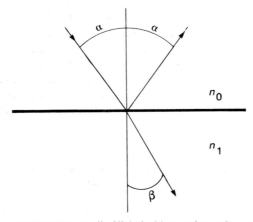

**Fig. 9.1**  Snell's law of refraction; pencil of light incident under angle $\alpha$ and refracted to angle $\beta$.

is totally reflected. If instead of a collimated beam, diffuse radiation is incident on an interface, integration over all angles and averaging is required to find the values for regular reflection of diffuse radiation (Kortüm, 1969).

$$R^* = 2 \int_0^{\pi/2} \sin \alpha \cos \alpha \, R(\alpha, n) \, d\alpha \tag{5}$$

where $R(\alpha, n)$ is formula (1) inserted into (2) with $n = n_0/n_1$.

For diffusely incident light from the medium with higher refractive index Judd (1942) gives a formula

$$R_i^* = 1 - \frac{T^*}{n^2} \tag{6}$$

where

$$T^* = 1 - R^* \tag{7}$$

$$T_i^* = 1 - R_i^*. \tag{8}$$

The crucial point is that $R_i^*$ contains significant contributions from totally reflected light, therefore $R_i^*$ is quite high. We will see that this is of great importance for the calculation of light gradients. The numerical value of $R_i^*$ is very sensitive to the assumption of true isotropic (diffuse) incidence of light, and may also vary if the interface is not perfectly smooth. However, Kottler (1960) and Mudgett and Richards (1971) have strongly advocated the use of the original formula. In Table 9.2 regular reflectances are listed for a range of biologically interesting refractive indices. The refractive index of water is 1.33, 1.40–1.45 covers the range of cell walls and many other living materials, 1.5 is the refractive index of glass and 1.55 is that of stratum corneum of skin. All these refractive indices refer to an object imbedded in a medium of refractive index 1.00 (air).

Another note is needed to clarify the difference between regular (or Fresnel) reflection of diffuse light as used above and diffuse reflection. Diffuse reflection occurs on a rough surface where the dimension of

**Table 9.2** Reflectance values and critical angle for selected values of the refractive index

| $n$ | 1.33 | 1.40 | 1.45 | 1.50 | 1.55 |
|---|---|---|---|---|---|
| $R'_\perp$ | 0.020 | 0.028 | 0.033 | 0.040 | 0.047 |
| $\beta_c$ | 48° 45′ | 45° 35′ | 43° 36′ | 41° 48′ | 40° 10′ |
| $R^*$ | 0.066 | 0.077 | 0.085 | 0.092 | 0.100 |
| $R_i^*$ | 0.472 | 0.529 | 0.565 | 0.596 | 0.625 |

roughness is in the order of the wavelength of incident light or slightly larger. For diffuse reflection, which is based empirically on the Lambert cosine law, the angular distribution is independent of the angle of incidence (Kortüm, 1969).

### 9.2.2. Absorption

Imagine a solution of pigment molecules. Let $c$ be the cross section (area) of the chromophor and $p$ the probability that a photon travelling through this area is absorbed. The effective absorption cross section of the molecule then is

$$\sigma_a = cp. \tag{9}$$

Now imagine a photon beam of cross section $C$ with a photon fluence rate of $I$ (mols quanta per second and per area). Let there be $n$ pigment molecules in a volume element $C\,dl$. Then

$$dI = -\sigma_a n d l I \tag{10}$$

$$\frac{dI}{I} = -\sigma_a n d l$$

and after integration

$$\ln \frac{I_t}{I_0} = -\sigma_a n l \tag{11}$$

or

$$I_t = I_0 e^{-\sigma_a n l}. \tag{12}$$

Setting

$$s_a = \sigma_a n \tag{13}$$

we have the absorption coefficient. Note that $\sigma_a = (m^2)$ and $s_a = (m^{-1})$. Changing bases yields the familiar Lambert–Beer–Boguer law

$$I_t = I_0 10^{-\varepsilon c l} \tag{14}$$

where $\varepsilon$ is the molar extinction coefficient, $c$ concentration and $l$ path length. Transformation gives, even more familiarly

$$\log \frac{I_0}{I_t} = A = \varepsilon c l. \tag{15}$$

$A$ is absorbance (formerly, but still not extinct, optical density), a pure number. The rule established by Equation (15), proportionality between absorbance and pigment concentration or optical pathlength, is well known and easily verified. For the sake of completeness, let's also state that in a mixture of $m$ pigments $i = 1 \ldots m$ absorbances are addititive:

$$A_{\mathrm{mix}\lambda_j} = \sum_{i=1}^{m} A_{i\lambda_j} = \sum_{i=1}^{m} \varepsilon_{i\lambda_j} c_i l. \tag{16}$$

Writing (and measuring) this for a set of $m$ wavelengths $\lambda_1 \ldots \lambda_m$ we get a system of $m$ linear equations in $\varepsilon_i$ or $c_i$ ($i = 1 \ldots m$), whatever is unknown, with a straightforward solution.

Now that we have been talking about light absorption in rather mathematical terms I feel that I owe the reader an explanation as to what light absorption means in a physicochemical interpretation. Absorption is the capture of a photon by a molecule (usually in its energetic ground state). As a consequence of this photon capture the molecule may be "lifted" in an excited state (energetically higher state owing to the energy of the absorbed photon), from which chemical reactions may start. The molecule may also return to its ground state by emitting a photon, usually of longer wavelength, i.e. less energy because of vibrational relaxation in the time span of $10^{-7}$ s between absorption and emission. This process is called fluorescence. Absorption of light by a molecule and energetic consequences are illustrated in the well-known Jablonski diagram (Moore, 1972). Light absorption without transition to an energetically higher state occurs when a photon is almost instantaneously re-emitted (i.e. within $10^{-15}$ s). This process is called scattering and will be dealt with in the next paragraph.

### 9.2.3.  Scattering

It is a familiar phenomenon to everybody that the rising or setting sun appears red while in the daytime it is usually yellowish. In contrast, the clear sky appears blue. Both colours, red and blue, are due to the same process, Rayleigh scattering of light by the molecules of the atmosphere. This type of scattering affects preferentially light of shorter wavelengths (blue) by deviating them from their original direction. Longer wavelengths, on the other hand, pass unscattered. Rayleigh scattering is characterized by the proportionality

$$\frac{I_s}{I_0} \propto \lambda^{-4} \tag{17}$$

where $I_s$ is the scattered intensity, $I_0$ the incident intensity, and applies only to particle diameters $d$ much smaller than $\lambda$, e.g. $d \leq \lambda/20$ (Moore, 1972).

Rayleigh scattering makes only a minor contribution to total scattering in biological tissue.

For larger particles, with sizes $d \geq \lambda$, the wavelength dependence vanishes (light scattered from a cloudy sky is white or grey). Single-particle scattering by spheres of a given refractive index is then described by the Mie theory (Kortüm, 1969). The Mie theory is the only general scattering theory available, therefore it has been frequently used in probably every field of research where light scattering is of interest. It should, however, be recognized that the Mie theory describes the rather special case of a single spherical particle and applications of the theory to particles of different shape or too closely spaced particles may yield erroneous results. Both Rayleigh and Mie scattering are examples of elastic or coherent scattering: the scattered photon has the same wavelength as the incident (unscattered) photons. Raman scattering is an example of inelastic scattering, where the photon takes up vibrational energy from the molecule hit or expends some energy to it. In this text we only deal with elastic scattering.

From Fig. 9.2 we see that light, after a scattering event, can travel in a direction more or less opposite to the incident beam. In a medium bounded by parallel planes (just for the ease of imagination) such photons may be re-emitted through the surface upon which the light beam was incident. This is called reflection or remission.

The following paragraphs expand some of the terminology associated with scattering, which may help to deal with advanced texts on scattering and radiative transfer. First consider a particle sitting in the origin of a coordinate system (Fig. 9.2). Let a thin light beam (a pencil of light) from direction $\alpha_i = (\theta_i, \varphi_i)$, in polar coordinates be scattered at the particle into the direction $\alpha_s = (\theta_s, \varphi_s)$. $f(\alpha_i, \alpha_s)$ is the relative probability that a photon is scattered from $\alpha_i$ to $\alpha_s$ and is called the scattering phase function. The simplest case of a phase function is isotropic scattering where all directions for the scattered photon are equally probable. Then

$$f = \omega_0 \tag{18}$$

and, for a general phase function $f$,

$$\frac{1}{4\pi} \int_{4\pi} f(\alpha_i, \alpha_s) \, d\alpha_s = \omega_0 \leq 1 \tag{19}$$

were $\omega_0$ is called the single scattering albedo (Ishimaru, 1978). $\omega_0$ is 1 for a pure scatterer (no absorption); this is also called conservative scattering. For randomly orientated scatterers a simplification is generally made, by assuming that the scattering phase function depends only on the angle $\psi$ between $\alpha_i$ and $\alpha_s$. By using $\cos \psi$ instead of $\psi$ we get

$$d\alpha = 2\pi \sin \psi d\psi \tag{20}$$

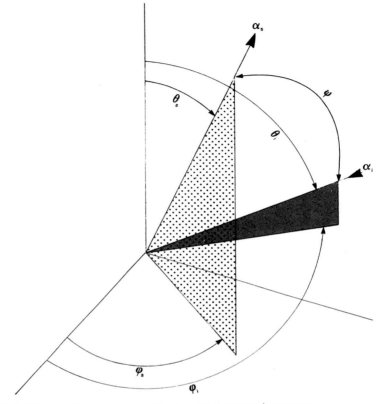

**Fig. 9.2**   The coordinate system used to describe the scattering process.

and

$$\omega_0 = \frac{1}{2} \int_{-1}^{1} f(\cos \psi) \, d(\cos \psi). \tag{21}$$

We can define a scattering cross section $\sigma_s$ in the same way as we described the absorption cross section (Equation (9)), in fact, $\sigma_a$ can be substituted by $\sigma_s$ in Equations (9) to (12) and the resulting equivalent to Equation (12) then describes how light is lost from an incident beam by outscattering. Defining the total cross section by

$$\sigma_t = \sigma_s + \sigma_a \tag{22}$$

and using $\sigma_t$ in Equations (9)–(13) an expression for the total loss from the incident beam follows

$$I_{t_c} = I_0 e^{-\sigma_t n l} \tag{23}$$

or with the definition

$$s_t = \sigma_t n \tag{24}$$

we obtain a more practical quantity

$$T_c = \frac{I_{t_c}}{I_0} = e^{-s_t l}. \tag{25}$$

There is a quite obvious relation between $\sigma_s$, $\sigma_a$ and $\omega_0$; since the latter is defined as the average fraction of photons scattered in photon–particle collisions (cf. Equation (19)) it follows that

$$\omega_0 = \frac{\sigma_s}{\sigma_s + \sigma_a} = \frac{\sigma_s}{\sigma_t}. \tag{26}$$

One more definition completes the collection of useful relations:

$$m = \frac{1}{s_t} \tag{27}$$

is defined as the mean free path of a photon, the path length in a scattering and or absorbing medium after which an incident beam is reduced to $1/e$ ~0.368 of its original intensity. $m$ is a useful value in Monte Carlo simulations of light distribution (Marchuk *et al.*, 1980).

## 9.3.  Measurable Quantities

This is meant to be a brief review of measuring techniques available for the determination of optical properties of biological tissue *in vivo*, or tissue removed or treated in some form.

The previous section on surface reflection, scattering and absorption has probably given an idea of how complex the pattern of light propagation in an absorbing and scattering tissue can be. We will first look into some of the integral optical properties of the tissue. Assume we have a slab of tissue bounded by parallel planes and unidirectionally illuminate it from one side. Then we have four major variables that can be observed: (i) **transmittance** $T$: the fraction of radiance leaving through the surface opposite to the illuminated one: (ii) **reflectance** $R$: the fraction of radiance re-emerging through the illuminated surface (a consequence of scattering); (iii) **surface reflection** $r$: as defined in Section 9.2.1, $r$ including $R_\perp$, $R_\alpha$, $R^*$; (iv) **absorption** $a$: the fraction of radiance converted to other forms of energy in the tissue.

Setting the incident light fluence rate to 1 we find

$$T + R + r + a = 1. \tag{28}$$

One note of caution: depending on definition you will find $r$ included in $R$ in some textbooks and publications.

Conventional spectrophotometers are not suitable to measure any of these quantities. Instead, an integrating sphere (Ulbricht sphere) is the required tool. Integrating sphere attachments are available for many commercial spectrophotometers. For purposes where high accuracy is not required, an integrating sphere can even be "home made" for little money. Typical integrating sphere measuring arrangements are shown in Figs 9.3–9.5; measurable or derivable quantities are listed in Table 9.3.

An integrating sphere is a hollow sphere (at least usually), for our purposes typically of 10–30 cm diameter, with three to five openings in the wall. It is covered on its inner surface with a diffusely and highly reflective coating, today in most cases $BaSO_4$, but formerly MgO was also used. Any light entering the sphere and reaching the inner sphere wall is multiply scattered and quite effectively and homogeneously redistributed on the

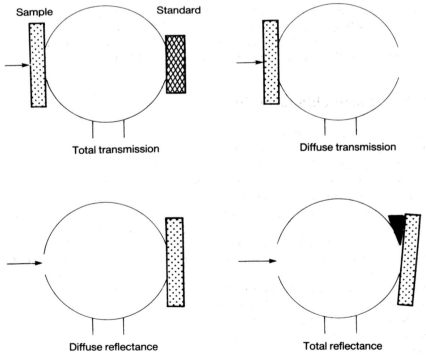

**Fig. 9.3** Substitution methods with integrating sphere. Light always enters from the left. For the reference measurement, the sample is replaced by the standard (reflectance) or the port is left blank (transmittance).

whole inner surface, i.e. the inner surface has a constant illuminance. This illuminance has a maximal value when all photons of a measuring beam reach the sphere wall, it is zero when no light reaches the wall and proportional in between. Now simply replace a small part of the surface by an appropriate photodetector. This is a photomultiplier in commercial integrating spheres; a photodiode might do in home-built instruments. One can now measure wall illuminance relative to standard values. Common

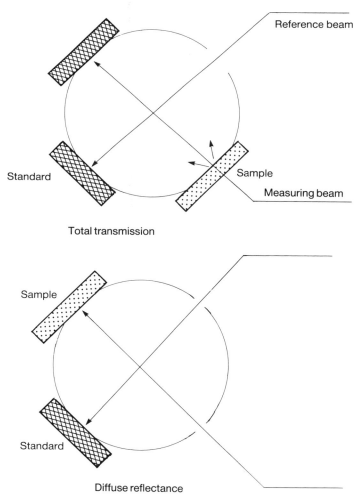

**Fig. 9.4**  Comparison methods with integrating sphere. Sample and standard remain in place all the time.

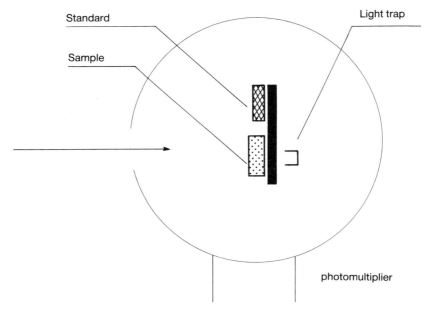

**Fig. 9.5** Absorbance measurement with integrating sphere. Sample and standard are moved into the beam alternately by means of a lever.

**Table 9.3** Quantities that can be determined directly or indirectly from integrating sphere measurements

| $R$ | Directly | Total reflectance |
|---|---|---|
| $T$ | Directly | Total transmittance |
| $R_d$ | Directly | Diffuse reflectance |
| $T_d$ | Directly | Diffuse transmittance |
| $a$ | Directly | Absorption |
| $T_c$ | $T - T_d$ | Collimated transmittance |
| $R'_\perp$ | $R - R_d$ | Specular reflectance |
| $a$ | $1 - R - T$ | Absorption |
| $A$ | $- \log T$ | Absorbance |

measuring geometries are shown in Figs 9.3–9.5. Methods can be classified in two groups; substitution method (Fig. 9.3) and comparison methods (Fig. 9.4). The substitution method needs a little more explanation. For every measured wavelength, after or before measuring wall illuminance with the sample in place, wall illuminance is also measured with a reference substituted for the sample. For transmittance measurements the reference is simply the blank entrance port; for reflectance measurements a reflectance standard is used, usually a pressed BaSO$_4$ tablet ($R \sim 0.96-0.98$). With the substitution method (Fig. 9.3) a systematic error in reflectance measurements is associated which comes from the fact that an almost perfectly white standard is replaced by a sample of usually much lower reflectance and therefore the average reflectance of the sphere wall decreases. This error mounts to as much as $-12\%$ for a sample window of 0.01 sphere surface (Jacques and Kuppenheim, 1955). It is eliminated in the comparison method, where sample and standard remain in place throughout the measurement and are alternately illuminated (in Fig 9.4 a double beam method is shown). Smaller systematic errors of around 1% result from the use of specular samples and standards.

Fig. 9.6 schematically depicts a method by which collimated transmittance $T_c$ can be measured and used to calculate $s_t$ (cf. Equation (25)). With a larger area photon detector in close proximity to the sample some fraction of scattered light can be measured too, but this is not a very reliable method unless one can assume a very sharply forward-peaked scattering distribution. Even then, the spatially dependent sensitivity of the detector limits accuracy.

The next set of conceivable measuring procedures used to quantify light

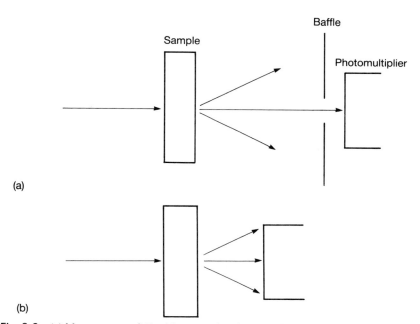

(a)

(b)

**Fig. 9.6**    (a) Measurement of $T_c$ with conventional spectrophotometer; (b) semi-quantitative measurement of $T$ with spectrophotometer.

distribution in biological tissue employs optical fibres. Optical fibres can be used for illumination purposes but also to collect light. They come in diameters from below 20 $\mu$m to millimeters, have acceptance angles (apertures) from around 20° to 120° and can be made from glass, quartz or plastic (not all specifications available with all materials, of course).

A good account of the use of light fibres in determination of the light distribution in plant tissue is given by Vogelmann and Björn (1984). In principle, thin optical fibres of 20–70 $\mu$m diameter were inserted into the tissue under a few representative angles and used to collect the light scattered in this direction or the collimated light (Fig. 9.7). To obtain the space irradiance at a given depth in the tissue, the contributions from fibres inserted under the sampling angles to the specified depth must then be extrapolated to cover the whole sphere around a point at the specified depth.

A different approach has been taken for animal tissue by Profio and Doiron (1987) and by Star et al. (1987); see also Marijnissen and Star, 1987). These researchers use quartz fibres (0.4 and 0.2 mm core diameter) coated with a scattering layer or bulb at one end. The tip is moved to different depths in a real or model tissue. Since the coated tips have a nearly isotropic response (to within 20–25%), so that photons from all directions $\alpha$ are collected with equal probability, a light detector attached to the other end of the fibre will yield a value proportional to local space irradiance at the diffusing tip. For both measuring protocols, in plant tissue with thin fibres and in animal tissue with isotropic tip fibres, a calibration is necessary to relate the detector signal to space irradiance. This calibration has to be done in a known light field in a clear medium of the same refractive index as the tissue medium (Marijnissen and Star, 1987). This is necessary because scattering properties of the diffusing tip depend on the refractive index of the surrounding medium.

The diffusing tip measurements require the insertion of a relatively large (in comparison to tissue extensions) scattering object into the tissue. How much the tissue optical properties are changed through this injury has not yet been tested; also the scattering tip itself may locally change the radiation regime. Nevertheless it can be stated that this approach gives results for the light gradient (gradient of space irradiance) quantitatively and qualitatively comparable to those of the Vogelmann and Björn method, as well as to those more analytical results derived in the next section.

One other method of interest uses fibre-optic bundles to measure tissue reflectance. From one end of the bundle the outer hull is removed and the bundle is separated in two smaller bundles (which receive new hulls), such that we have a bifurcate arrangement. By illuminating one of the smaller bundles with a light source, light can be guided to the end of the large

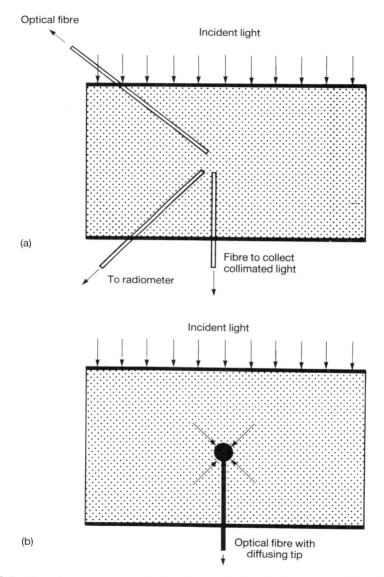

**Fig. 9.7** Measuring geometries with light fibres: (a) after Vogelmann and Björn (1984); (b) following Star *et al.* (1987) (fibre and sample not drawn to scale).

bundle, which can be positioned at the surface of tissue. Through the fibres of the other arm, which also end in the large bundle at the surface of the tissue, backscattered light is then collected and guided to a detector (Dawson *et al.*, 1980). This method can also be applied to fluorescence measurements with the same rationale (Chow *et al.*, 1981). Although being rather convenient in experimental terms, this approach is only semi-quantitative because of the limited aperture of the fibres.

Before closing this section I would like to mention briefly one method, photoacoustic spectroscopy, that cannot be covered in detail here but that has the potency to complement methods described in this chapter. With photoacoustic spectroscopy one can (in principle) measure absorption gradients in scattering and non-scattering tissue. A recent review of the method, with a focus on photosynthesis research, has been given by Buschmann *et al.* (1984) and a more general coverage is provided by Rosencwaig (1980).

## 9.4. Models for Photon Fluxes in Tissue

### 9.4.1. Kubelka–Munk theory and simple extensions

Let us start with a heuristic or phenomenological model that can be formulated without using physical description of scattering or absorption. The Kubelka–Munk approach is the simplest model that includes scattering and absorption. Of course Beer's law does not apply to scattering media.

The model is applied to a plane parallel (or semi-infinite) medium and involves two diffuse fluxes $I$ and $J$ travelling in the positive or negative $x$ directions (Fig. 9.8). Each flux compromises all photons travelling within a solid angle of $2\pi$ (Kubelka, 1948). A number of assumptions underlie the model. The scattering layer extends infinitely in the directions perpendicular to the beam to avoid edge effects. The matrix, in which scatterers and absorbers are imbedded, has the same refractive index as the surrounding medium. The scattering layer is macrohomogeneous, and the incident photon flux is diffuse. Moreover, let us assume a beam of infinite sideways extent to compensate for outscattering from the beam cross section.

Then the flux $I$ loses photons by absorption and by scattering to $J$, and gains photons from scattering by $J$. With the phenomenological scattering coefficient $S$ and absorption coefficient $K$ (in $mm^{-1}$) we can write a system of two differential equations describing the two fluxes.

$$\frac{dI}{dx} = -KI - SI + SJ \qquad (29)$$

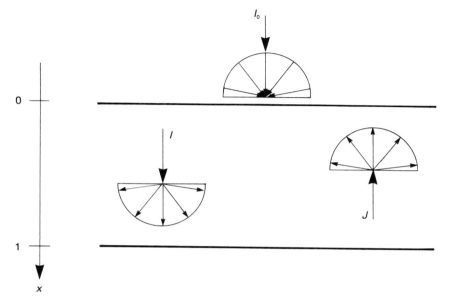

**Fig. 9.8**  Light fluxes in the Kubelka–Munk theory.

$$\frac{\mathrm{d}J}{-\,\mathrm{d}x} = -KJ - SJ + SI. \tag{30}$$

$K$ and $S$ are phenomenological constants (depending on $\lambda$ of course) inasmuch as they have no immediate physical meaning attached to them. The $-\,\mathrm{d}x$ accounts for the direction of the $J$ flux which is opposite to that of the $I$ flux. Solving the system with boundary conditions.

$$I(0) = I_0 \qquad J(0) = R$$

$$I(l) = T \qquad J(l) = 0$$

with $T$ and $R$ diffuse transmittance and reflectance—that is what we need the integrating sphere for—one gets a set of useful equations (Kubelka, 1948; Judd and Wyszecki, 1963; Wendlandt and Hecht, 1966; Kortüm, 1969)

$$S = \frac{1}{bl}\left[ \coth^{-1}\left( \frac{1 - aR}{bR} \right) \right] \tag{31}$$

$$K = S(a - 1) \tag{32}$$

$$a = \frac{1 + R^2 - T^2}{2R} = \frac{S + K}{S} \tag{33}$$

$$b = \sqrt{a^2 - 1} \tag{34}$$

$$R_\infty = a - b \tag{35}$$

$$F(R_\infty) = \frac{(1 - R_\infty)^2}{2R_\infty} = \frac{K}{S} \tag{36}$$

$$I(x) = I_0 \left[ \frac{R - a}{b} \sinh (Sbx) + \cosh (Sbx) \right] \tag{37}$$

$$J(x) = I_0 \left[ \frac{aR - 1}{b} \sinh (Sbx) + R \cosh (Sbx) \right] \tag{38}$$

$$w(x) = I(x) + J(x)$$
$$= I_0 \left[ (1 + R) \cosh (Sbx) - \frac{a + 1}{b} (1 - R) \sinh (Sbx) \right]. \tag{39}$$

Equations (29) and (30) can also be solved with different boundary conditions taking into account a reflective backing which is certainly useful in some applications.

In daily practice, the set of conditions under which the Kubelka–Munk theory is strictly valid is hardly ever fulfilled. The first and most frequent violation is given by the use of collimated incident light. The justification for this lies in the assumption, that the beam is being diffused "rapidly enough". In fact model calculations have shown that, assuming Mie scattering, collimated incident light is diffuse after two to eight scattering events (Kortüm, 1969). Results obtained by Vogelmann and Björn back the use of collimated light experimentally, at least for relatively thick samples. The second violation results from the refractive index of the tissue being higher than that of the surrounding medium. This has a serious consequence for calculations of fluence rate gradients. As we have seen, the reflectance at a step down of refractive index from 1.45 to 1.0 is 56.5% for diffuse light (Table 9.2). Thus photons entering a scattering tissue of higher refractive index in a collimated beam and then scattered are quite effectively hindered from leaving the tissue, which makes it act as a light trap (Vogelmann and Björn, 1984; Seyfried and Schäfer, 1985) and has dramatic effects on the light gradient (Seyfried and Fukshansky, 1983). The problem with specular reflective boundaries arises first when one tries to determine $K$ and $S$ from measured $R$ and $T$ values. These values do not represent the scattering tissue but are changed by the effects of surface reflection. A remedy to this problem can be given by mathematically "removing" the surface reflection, which works when the refractive index of the tissue is known (Seyfried *et al.*, 1983). Another procedure to solve the same problem is treated in great detail by Fukshansky-Kazarinova *et al.* (1986); these authors also include an extension of the analysis to a two-layered object, where the Kubelka–Munk theory is not naturally applicable (layered objects are not macrohomogeneous).

Representative graphs of light gradients in scattering and absorbing biological tissue of higher refractive index than the surrounding medium can be found in Fukshansky-Kazarinova *et al.* (1986), Seyfried and Schäfer (1985), Seyfried and Fukshansky (1983), Vogelmann and Björn (1984), Star *et al.* (1988) and Marijnissen and Star (1987). Figure 9.9 gives two examples of light gradients for single-layer models following the Kubelka–Munk theory with surface reflection. There it is also illustrated how the wrong light gradients are obtained when surface reflection is neglected. There are two possibilities to do it wrongly: first, take a scattering object, measure $R$ and $T$, then calculate $S$ and $K$ as if there were no reflecting boundary. This gives the dotted line in Fig. 9.9. Second, use the true coefficients, but calculate light gradients neglecting surface reflection. This gives the dashed line in Fig. 9.9. Note that the correct light gradient can give values much higher than $I_0$. This demonstrates backscattering and light trapping; it does not violate the law of energy conservation. The two small arrows in this figure indicate the values of $1 + R$ and $T$ for the highly scattering system. This should make it quite clear that $R$ and $T$ do not provide direct information on the value of the light gradient anywhere inside the object. This information has to be calculated from the two values and other parameters first.

As has been stated, the Kubelka–Munk theory often has to be used in situations where it is formally more or less outside its range of validity. Therefore we cannot expect the Kubelka–Munk theory to yield exact results. Still the calculations can be quite lengthy. Therefore an approximation of the light gradient can often be used that avoids the intermediate step of $K$ and $S$ calculation (Seyfried and Schäfer, 1985). This also helps to get over the psychological barrier of being concerned about the impact of tissue optical properties on photobiological experiments. When $R + T \leq 0.90$ (as measured with an integrating sphere, such that $R' = 1 - T'$ is not detected) we can approximate the light gradient

$$w(x) = w(0)e^{-qx} \qquad (40)$$

i.e. an exponential function through $w(0)$ and $w(l)$ with

$$q = \frac{1}{l} \ln \frac{w(0)}{w(l)} \qquad (41)$$

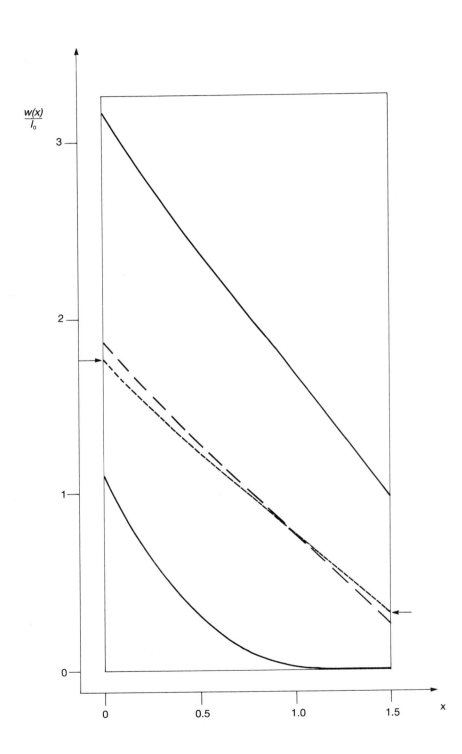

and

$$w(0) = T' + \frac{R}{T_i^*} (1 + R_i^*) \qquad (42)$$

$$w(l) = \frac{T}{T_i^*} (1 + R_i^*). \qquad (43)$$

For weakly absorbing objects with $R + T > 0.9$ a linear approximation through $w(0)$ and $w(l)$ is more appropriate. Errors in comparison to ideal Kubelka–Munk theory are below 10%.

As we are already diverging a little from the Kubelka–Munk theory let us just look into how the formulae (42) and (43) have been obtained. The idea is that $R$ results from a flux $J$ propagating upwards (Fig. 9.8) through the partially reflecting boundary

$$R = J(0)T_i^* \qquad (44)$$

so that

$$J(0) = \frac{R}{T_i^*}. \qquad (45)$$

$J$ also gives rise to part of the downward travelling flux by virtue of being reflected at the boundary. The rest of the $I$ flux is from incident light (here assumed to be collimated) so that

$$I(0) = T' + \frac{RR_i^*}{T_i^*} \qquad (46)$$

and

$$I(0) + J(0) = w(0). \qquad (47)$$

Now that we have seen that addition of surface reflection makes an object formally multilayered (the surfaces being pseudolayers), I would like to add another useful point of view. There is an easy way to calculate the total reflectance and transmittance of a multilayered object when transmittance and reflectance of single layers are known. Let $R_{1u}$ and $R_{1l}$ designate the reflectances of layer 1 (so it need not be homogeneous, in which case the reflectances were equal), and $T$ its transmittance (which is independent of orientation, see Kubelka (1948) for proof and Equation (50) below), and analogously for layer 2. Put the two layers together at sides 1l and 2u (l and

---

**Fig. 9.9** Fluence rate gradients: bottom line for $K_1 = 3$ and $S_1 = 1$; top line for $K_2 = 0.02$ and $S_2 = 2$; both with correct account for specular reflection ($n = 1.425$); dotted line: wrong light gradient calculated from $K_2'$ and $S_2'$ obtained by neglecting surface reflection; dashed line: wrong light gradient calculated with correct $K_2$ and $S_2$ neglecting surface reflection.

u stand for lower and upper, as you have probably noted). We then have (Stokes, 1862; Wendlandt and Hecht, 1966; Kortüm, 1969; Fukshansky, 1981; Seyfried *et al.*, 1983; Gross *et al.*, 1984, and many more)

$$R_{1,2} = R_{1u} + \frac{T_1^2 R_{2u}}{1 - R_{1l}R_{2u}} \tag{48}$$

$$R_{2,1} = R_{2l} + \frac{T_2^2 R_{2l}}{1 - R_{1l}R_{2u}} \tag{49}$$

$$T_{1,2} = T_{2,1} = \frac{T_1 T_2}{1 - R_{1l}R_{2u}} \tag{50}$$

and I shall not give the full derivation (except to say that it is based on geometric series and involves adding the multiple up- and down-scatter of photons between the two layers), because it has already been derived in so many books and papers. It should be noted that these three equations can be used iteratively, i.e. after fusion of two layers just proceed to add a third layer to the already existing non-homogeneous double layer and so on.

### 9.4.2. Three- and four-flux models

The major criticism towards the application of the Kubelka–Munk theory arises from the requirement of diffusely incident light. This is hardly ever used. It has therefore been attempted to include a third, collimated flux $i$ that loses photons by absorption or by scattering to $I$ or $J$. Scattering to $I$ and $J$ may be different.

Schuster (1905) and Völz (1962, 1964) have developed four-flux models with an additional collimated flux $j$ directed opposite to $i$. We here follow the quite readable paper by Mudgett and Richards (1971); in our notation we have

$$\frac{di}{dx} = -(s_a + S_1 + S_2)i \tag{51}$$

$$\frac{dI}{dx} = S_1 i - (K + S)I + SJ + S_2 j \tag{52}$$

$$\frac{dJ}{-dx} = S_2 i + SI - (K + S)J + S_1 j \tag{53}$$

$$\frac{dj}{-dx} = -(s_a + S_1 + S_2)j. \tag{54}$$

In this set of equations the assumption is made that scattering from $I$ and $J$ into the collimated beams is negligible. Leaving out all collimated fluxes reduces the system to the Kubelka–Munk theory.

The four-flux model leaves us with a set of five unknowns, $S_1$, $S_2$, $S$, $K$ and $s_a$. We may replace

$$K = 2s_a \qquad (55)$$

because diffuse light has an average pathlength just twice that of collimated light. Using a set of four independent measurements yields boundary conditions that allow solving for the remaining variables. Van Gemert et al. (1987), in a slightly different model, suggest measurements with diffusely incident and collimated light and thus obtain three independent measuring values $R$, $T$ and $T_c$. Still, they have to use a heuristic assumption to relate a fourth term. A complete and practicable solution to the four-flux or three-flux problem is not known to the author. Note that the $j$ flux does not contain any extra information, so leaving this one out (which is justified if it is very small) would not improve on the situation. We will come back to the Kubelka–Munk theory and four-flux models from a more physical point of view in the next section on radiative transfer.

## 9.5.  The Radiative Transfer Equation and its Relation to Two- and Four-flux Models

The equation of radiative transfer (Chandrasekhar, 1960) describes the changes in number of photons travelling in direction $\alpha$ through a given point $\mathbf{r}$:

$$\alpha\ \mathrm{grad}I(\mathbf{r}, \alpha) = -\ \sigma_t I(\mathbf{r}, \alpha) + \sigma_s \int_{4\pi} I(\mathbf{r}, \alpha_i) f(\alpha_i, \alpha)\ \mathrm{d}\alpha_i. \qquad (56)$$

$-\ \sigma_t I(\mathbf{r}, \alpha)$ accounts for losses through absorption and scattering out of the $\alpha$ direction. The integral term represents scattering from all directions $\alpha_i$ into $\alpha$, with the phase function $f$ introduced in Equation (19). It is justified to assume that the scattering phase function depends only on the angle $\psi$ between $\alpha_i$ and $\alpha$ (compare Equations (20) and (21)). Observing a beam travelling along the $x$-axis we have (with $\mu = \cos\psi$)

$$\mu\ \frac{\mathrm{d}I(x, \mu)}{\mathrm{d}x} = -\ \sigma_t(x, \mu) + \frac{\sigma_t}{2} \int_{-1}^{1} I(x, \mu) f(\mu)\ \mathrm{d}\mu. \qquad (57)$$

The problems and merits of the radiative transfer equation lie in the fact that the scattering phase function is an explicit part of the equation. Analytical treatment of this integro-differential equation is not usually possible; instead, numerical methods have to be used to solve for $I(x, \mu)$ and $I(x)$.

While the literature abides with approximations, expansions of the phase function in Legendre polynomials and Monte Carlo simulations, I will only mention here a few results that bear on two- and four-flux approximations.

It is straightforward to write for the light scattered in a forward direction

$$\beta = \frac{1}{2\omega_0} \int_0^1 f(\mu) \, d\mu \tag{58}$$

and $1 - \beta$ for the light scattered backwards. Of course one has to know the phase function to make use of this information. Another way of taking differences between forward and backscattering into account is found by building an artificial phase function with an asymmetry (or anisotropy) parameter $g$. A quite frequently used phase function is given by

$$f(\mu) = \frac{1 - g^2}{(1 + g^2 - 2g\mu)^{3/2}}, \tag{59}$$

the Henyey–Greenstein phase function. This function is adjusted such that

$$\int_{-1}^1 f(\mu)\mu \, d\mu = g = \bar{\mu}. \tag{60}$$

Just for comparison, let us also have a look at the phase function belonging to Rayleigh scattering:

$$f(\mu) = \tfrac{3}{4} (1 + \mu^2), \tag{61}$$

which is symmetric in forward and backward direction.

The general belief is that scattering in biological tissue is forward peaked (Diffey, 1983; Latimer and Noh, 1987) with $\beta$ somewhere around 0.84 (Diffey, 1983). One relation that is important for us is the one between the Kubelka–Munk theory $K$ and $S$ and $\sigma_a$ and $\sigma_s$. They are *not* the same. To find out how they compare we make three more definitions following Meador and Weaver (1979)

$$\gamma = 1 - \omega_0 \qquad \eta = \frac{s_a}{K} \qquad G = \frac{s_s}{S}.$$

These values are meant for isotropic scattering; see van Gemert and Star (1987) for the anisotropic case. Meador and Weaver find

$$2\eta = 1 + 0.5\gamma, \tag{62}$$

i.e. a deviation from the expected $2\eta = 1$ relationship with increasing absorption. The scattering coefficient is a little more complicated:

$$\frac{9\sigma}{4} = 3 + 1.9\gamma. \tag{63}$$

Multiflux calculations by Mudgett and Richards (1971) showed that predictions from multiflux models and the Kubelka–Munk theory start to deviate above $\sigma_s/\sigma_a$ values of 0.01, thus supporting Meador and Weaver's results.

Apart from the applications cited and a number of related publications little additional help is available for practical purposes from the radiative transfer theory. This is a field where some good and practicable new ideas are most welcome.

## 9.6.  More Effects to Consider

### 9.6.1.  Optical pathlength increase

When a light beam interacts with a scattering medium some of the light is scattered multiply before being transmitted and may (possibly) be measured as transmittance. We have seen earlier that absorbance $A$ is related to transmittance $T$ and pathlength (Equation (15). Consequently $A$ increases with scattering, because the optical pathlength (not the geometrical pathlength) increases. In fact, apparent absorbance increases even more because some of the forward scattered light misses the detector, as does all of the backward scattered light. There have been attempts to quantify the effect (Butler, 1962) by defining a pathlength increase factor. This factor is a constant that applies to a given object under certain measuring conditions; it is not a material constant, but depends also in a rather complex way on object thickness and refractive index. One practical application lies in the amplification of small absorbance differences after addition of a scatterer to the solution. This is, however, paid for by a decreased signal-to-noise ratio in the transmittance measurement. It is anyway preferable not to use transmittance measurements in such a case but an integrating sphere and to make calculations with $1 - a$ instead of $T$.

Different scattering properties of tissues of different age or hydration make comparison of pigment absorption difficult. Here it may help to grind up the tissue and mix it with a strong scatterer (MgO has been used) such that the differences of scattering between tissues become negligible.

The pathlength increase discussion, in my view, is a more or less artificial one because it relates to the application of a method (transmittance measurement) to a problem where the method is unsuitable.

### 9.6.2.  Sieve effect

Another complication arises when we use pigment with inhomogeneous pigment distribution in spectroscopy methods. This refers to the "little packages" of chlorophyll from one of the introductory examples. Any such local concentration of pigments tends to decrease absorbance as compared to a homogeneous solution of the same amount of pigment (Fukshansky, 1978; Gross *et al.*, 1984) and to flatten the absorbance spectrum (Duysens, 1956). A more complete theory of the sieve effect as it is also influenced by scattering (that is where things get complicated) is given by Fukshansky (1987) along with some practical application methods.

### 9.6.3. Fluorescence

Up to now we have focused our interest on effects of light that come from outside. Upon absorption, this light can also give rise to light that comes from inside, i.e. fluorescence. Fluorescence can be accounted for in the radiative transfer equation by introducing an additional source term, but we would have to rewrite the whole equation to take care of the changes in frequency (or wavelength) between absorbed and emitted light. The difficulty is that the local fluorescence production depends on the local light regime and on the local concentration of fluorescing pigment. Problems with fluorescence arise either when you simply plan to measure transmittance and are unaware of possible contributions from fluorescence to your transmittance value, or when you use fluorescence as a tool to collect information, e.g. on absorption or energy transfer processes. Approaches to this problem are available (Fukshansky and Kazarinova, 1980; Lork and Fukshansky, 1985; Seely, 1987), but a lot more work needs to be done.

### 9.6.4  Particular surfaces of natural tissue

When working with hairless mice and tissue from specially selected plant species one tends to forget that many living organisms are equipped with special surface structures that may either protect them from excessive radiation or enable them to collect light more efficiently. These structures can be hairs as found on many animals and plants (some desert and alpine plants have thick white hair coverings), thin waxy layers in plants and lens-like structures on leaf surfaces. This collection is quite interesting because three different effects are involved. Hairy layers may probably be accounted for in terms of a scattering layer (see above); waxy layers are more a problem of wave optics and interference; and lenses relate to geometric optics. Little effort has so far been spent to analyse the consequences of such surface structures for light distribution.

## 9.7. Light Gradients and Action Spectroscopy

The two basic quantities to be known in action spectroscopy are the applied photon fluence and the response measured as a consequence. In applications with living tissue the photon fluence depends on penetration depth. Also the ability of the tissue to respond to light may depend on the location of the photosensory pigment. There have been a few attempts to put action spectroscopy on a basis where it could cope with light gradients, but these approaches are either not readily applicable in the form presented (Hartmann and Cohnen-Unser, 1972) or restricted to a special case (Kazarinova-Fukshansky et al., 1985).

It seems to be necessary to take different approaches depending on the interesting pigment for a particular study. If we have a mass pigment (say chlorophyll) that by itself strongly influences the light gradient, the measured response $r$ will be

$$r \propto a \int_0^1 \alpha_p(x) w(x) \, dx \tag{64}$$

if $\alpha_p$ is the contribution of the pigment to the local absorption probability, and $a = 1 - R - T$. There the light gradient is a measure for the distribution of the measured effect, provided (we have assumed this implicitly) the pigment concentration is constant.

For a pigment with only a small contribution to total absorption we can simplify to

$$r \propto \int_0^1 p(x) w(x) \, dx \tag{65}$$

where $p(x)$ is the distribution of the pigment. Here we only have to know the distribution of the sensory pigment itself, but not the distribution of all other major pigments. For a homogeneously distributed pigment we simply have

$$r \propto p \int_0^1 w(x) \, dx. \tag{66}$$

This integral can be considerably larger than 1 in highly scattering tissue, or quite small, e.g. for blue light in most living tissues (see Fig. 9.9).

We often find pigments that act as a screen to protect parts of the tissue from excessive light flux. It seems logical that such a screen should be located between light source and protected tissue. In reality such a screen could, however, also be located behind the tissue as can be seen from the two equations above. They state that the response (for example of sensitive tissue to light) depends on all of the tissue, not only on the layers above a given site of action. A screen is more effective, though, if it is between light source and sensitive tissue. As an interesting point one should note that action spectra depend on the orientation of the observed tissue, but absorbance spectra do not, because $T$ does not. The absorption $a$ of a tissue, on the other hand, again depends on orientation, as does $R$.

As we see, it is not too difficult to understand basic light–tissue interactions, but we still have a long way to go before we can solve the more advanced problems. The concepts and models available at this time, however, often enable the researcher in photobiology or photomedicine to obtain a good quantitative and qualitative estimate of light–tissue interaction.

# References

Baskin, T. I. & Iino, M. (1987). An action spectrum in the blue and ultraviolet for phototropism in alfalfa. *Photochem. Photobiol.* **46**, 127–136.

Buschmann, K., Prehn, H. & Lichtenthaler, H. (1984). Photoacoustic spectroscopy (PAS) and its application in photosynthesis research. *Photosynth. Res.* **5**, 24–46.

Butler, W. L. (1962). Absorption of light by turbid materials. *J. Opt. Soc. Am.* **52**, 292–299.

Chandrasekhar, S. (1960). "Radiative Transfer". Dover, New York.

Chow, W. S., Telfer, A., Chapman, D. J. & Barber, J. (1981). State 1-State 2 transition in leaves and its association with ATP-induced chlorophyll fluorescence quenching. *Biochim. Biophys. Acta* **638**, 60–68.

Dawson, J. B., Barker, D. J., Ellis, D. J., Grassam, F., Cotterill, J. A., Fister, G. W. & Feather, J. W. (1980). A theoretical and experimental study of light absorption and scattering by *in vivo* skin. *Phys. Med. Biol.* **25**, 695–709.

Diffey, B. L. (1983). A mathematical model for ultraviolet optics of skin. *Phys. Med. Biol.* **28**, 647–657.

Dougherty, T. J., Boyle, D. G. & Weishaupt, K. R. (1982). Photoradiation therapy of human tumours. *In* "The Science of Photomedicine" (J. D. Regan & J. A. Parrish, Eds), pp. 625–638. Plenum, New York.

Duysens, L. N. M. (1956). The flattening of the absorption spectrum of suspensions as compared to that of solutions. *Biochim. Biophys. Acta* **19**, 1–12.

Fukshansky, L. (1978). On the theory of light absorption in nonhomogeneous objects: the sieve effect in one-component systems. *J. Math. Biol.* **6**, 177–196.

Fukshansky, L. (1981). Optical properties of plant tissue. *In* "Plants and the Daylight Spectrum" (H. Smith, Ed.), pp. 21–40. Academic, London.

Fukshansky, L. (1987). Absorption statistics in turbid media. *J. Quant. Spectr. Radiat. Transfer* **38**, 389–406.

Fukshansky, L. & Kazarinova, N. (1980). Extension of the Kubelka-Munk theory of light propagation in intensely scattering materials to fluorescent media. *J. Opt. Soc. Am.* **70**, 1101–1111.

Fukshansky-Kazarinova, N., Lork, W., Schäfer, E. & Fukshansky, L. (1986). Photon flux gradients in layered turbid media; application to biological tissues. *Appl. Optics* **25**, 780–788.

Gausman, H. W. (1985). Plant Leaf Optical Properties in Visible and Near-infrared Light. Texas Tech. Press Lubbock, Texas.

van Gemert, M. J. C. & Star, W. M. (1987). Relations between the Kubelka-Munk and the transport equation models for anisotropic scattering. *Lasers Life Sci.* **1**, 287–298.

van Gemert, M. J. C., Welch, A. J., Star, W. M., Motamedi, M. & Cheong, W.-F. (1987). Tissue optics for a slab geometry in the diffusion approximation. *Lasers Med. Sci.* **2**, 295–302.

Govindjee, Amesz, J. & Fork, D. C. (Eds) (1986). "Light Emission by Plants and Bacteria". Academic, Orlando.

Groenhuis, R. A. J., ten Bosch, J. J. & Ferwerda, H. A. (1983). Scattering and absorption of turbid materials determined from reflection measurements. 2: Measuring method and calibration. *Appl. Optics* **22**, 2463–2467.

Gross, J., Seyfried, M., Fukshansky, L. & Schäfer, E. (1984) *In vivo* spectrophotometry. *In* "Techniques in Photomorphogenesis" (H. Smith & M. G. Holmes, Eds), pp. 131–157. Academic, London.

Grum, F. & Becherer, R. J. (1979). "Optical Radiation Measurements 1". Academic, New York.

Harm, W. (1980) "Biological Effects of Ultraviolet Radiation". Cambridge University Press, Cambridge.

Hartmann, K. M. (1983). Action spectroscopy. *In* "Biophysics" (W. Hoppe, W. Lohmann, H. Markl & H. Ziegler, Eds), pp. 115–144. Springer, Berlin.

Hartmann, K. M. & Cohnen-Unser, I. (1972). Analytical Action Spectroscopy with living systems: Photochemical aspects and attenuance. *Ber. Deutsch. Bot. Ges.* **85**, 481–551.

Ishimaru, A. (1978). "Wave Propagation and Scattering in Random Media", Vols I and II. Academic, New York.

Jacques, A. J. & Kuppenheim, H. F. (1955). Theory of the integrating sphere. *J. Opt. Soc. Am.* **45**, 460–470.

Judd, D. B. (1942). Fresnel Reflection of Diffusely Incident Light. *J. Res. Natl. Bur. Std.* **29**, 329–343.

Judd, D. B. & Wyszecki, G. (1963) "Color in Business, Science and Industry". Wiley, New York.

Kazarinova-Fukshansky, N., Seyfried, M. & Schäfer, E. (1985). Distortion of action spectra in photomorphogenesis by light gradients within the plant tissue. *Photochem. Photobiol.* **41**, 689–702.

Kortüm, G. (1969). "Reflectance Spectroscopy". Springer, Berlin.

Kottler, F. (1960). Turbid media with plane parallel surfaces. *J. Opt. Soc. Am.* **50**, 483–490.

Kubelka, P. (1948). New contributions to the optics of intensely light scattering materials, Part I. *J. Opt. Soc. Am.* **38**, 448–457.

Latimer, P. & Noh, S. J. (1987). Light propagation in moderately dense particle systems: a reexamination of the Kubelka-Munk theory. *Appl. Optics* **26**, 514–523.

Lork, W. & Fukshansky, L. (1985). The influence of chlorophyll-fluorescence on the light gradients and the phytochrome state in a green model leaf under natural conditions. *Plant Cell Envir.* **8**, 33–39.

Lübbers, D. W. & Wodick, R. (1975). Absolute reflection photometry applied to the measurement of capillary oxyhemoglobin saturation of the skin in man. *In.* "Oxygen Measurement in Biology and Medicine" (J. P. Payne & D. W. Hill, Eds), pp. 85–110. Butterworths, London.

Mandoli, D. F. & Briggs, W. R. (1982). The photoreceptive sites and the function of tissue light piping on photomorphogenesis of etiolated oat seedlings. *Plant Cell Envir.* **5**, 137–145.

Marchuk, G. I., Mikhailov, G. A., Nazaraliev, M. A., Darbinjan, R. A., Kargin, B. A. & Elepov, B. S. (1980). "The Monte Carlo Method in Atmospheric Optics". Springer, Berlin.

Marijnissen, J. P. A. & Star, W. M. (1987). Quantitative light dosimetry *in vitro* and *in vivo*. *Lasers Med. Sci.* **2**, 235–242.

Meador, W. E. and Weaver, W. R. (1979) Diffusion approximation for large absorption in radiative transfer. *Appl. Optics* **18**, 1204–1208.

Moore, W. J. (1972) "Physical Chemistry". Longman, London.

Mudgett, P. S. & Richards (1971). Multiple scattering calculations for technology. *Appl. Optics* **10**, 1485–1502.

Parsons, A., Macleod, K., Firn, R. D. & Digby, J. (1984). Light gradients in shoots subjected to unilateral illumination – implications for phototropism. *Plant Cell Envir.* **7**, 325–332.

Profio, A. E. & Doiron, D. R. (1987). Transport of light in tissue in photodynamic therapy. *Photochem. Photobiol.* **46**, 591–599.

Regan, J. D. & Parrish, J. A. (Eds) (1982). "The Science of Photomedicine". Plenum, New York.

Rosencwaig, A. (1980). "Photoacoustics and Photoacoustic Spectroscopy". Wiley, New York.

Schäfer, E., Fukshansky, L. & Shropshire, W. (1983). Action spectroscopy. *In* "Photomorphogenesis. Encyclopedia of Plant Physiology", N.S., Vol. 16A (W. Shropshire & H. Mohr, Eds), pp. 31–68. Springer, Berlin.

Schuster, A. (1905). Radiation through a foggy atmosphere. *Astrophys. J.* **21**, 1–21.

Seely, G. R. (1987). Transport of fluorescence through highly scattering media. *Biophys. J.* **52**, 311–316.

Seyfried, M. & Fukshansky, L. (1983). Light gradients in plant tissue. *Appl. Optics* **22**, 1402–1408.

Seyfried, M. & Schäfer, E. (1983). Changes in the optical properties of cotyledons of *Cucurbita pepo* during the first seven days of their development. *Plant Cell Envir.* **6**, 633–640.

Seyfried, M. & Schäfer, E. (1985). Phytochrome macrodistribution, local photoconversion and internal photon fluence rate for *Cucurbita pepo* L. cotyledons. *Photochem. Photobiol.* **42**, 309–318.

Seyfried, M., Fukshanky, L. & Schäfer, E. (1983). Correcting remission and transmission spectra of plant tissue measured in glass cuvettes: A technique. *Appl. Optics* **22**, 492–496.

Star, W. M., Marijnissen, H. P. A., Jansen, H., Keijzer, M. & van Gemert, M. J. C. (1987). Light dosimetry for photodynamic therapy by whole bladder wall irradiation. *Photochem. Photobiol.* **46**, 619–624.

Star, W. M., Marijnissen, J. P. A. & van Gemert, M. J. C. (1988). Light dosimetry in optical phantoms and in tissues: I. Multiple flux and transport theory. *Phys. Med. Biol.* **33**, 437–454.

Steinhardt, A. R., Shropshire, W. & Fukshansky, L. (1987). Invariant properties of absorption profiles in sporangiophores of *Phycomyces blakesleeanus* under balancing bilateral illumination. *Photochem. Photobiol.* **45**, 515–523.

Stokes, G. S. (1862). On the intensity of light reflected from or transmitted through a pile of plates. *Proc. R. Soc. London* **11**, 545–557.

Terashima, I. & Saeki, T. (1983). Light environment within a leaf. I. Optical properties of paradermal sections of *Camellia* leaves with special reference to differences in the optical properties of palisade and spongy tissues. *Plant Cell Physiol.* **24**, 1493–1501.

Völz, H. G. (1962). Ein Beitrag zur phänomenologischen Theorie lichtstreuender und absorbierender Medien, VI. FATIPEC Kongress 1962, pp. 98–103. Verlag Chemie, Weinheim.

Völz, H. G. (1964). Ein Beitrag zur phänomenologischen Theorie lichtstreuender und absorbierender Medien. Teil II: Möglichkeiten zur experimentellen Bestimmung der Konstanten. VII. FATIPEC Kongress 1964, pp. 194–201. Verlag Chemie, Weinheim.

Vogelmann, T. C. & Björn, L. O. (1984). Measurement of light gradients and spectral regime in plant tissue with fiber optic probe. *Physiol. Plant.* **60**, 361–368.

Weis, E. (1985). Chlorophyll fluorscence at 77 K intact leaves: Characterisation of a technique to eliminate artefacts related to self-absorption. *Photosynth. Res.* **6**, 73–86.

Wendlandt, W. W. M. & Hecht, H. G. (1966) "Reflectance Spectroscopy". Wiley, New York.

# Index